EXCEL®
STUDENT LABORATORY MANUAL AND WORKBOOK

BEVERLY DRETZKE
University of Minnesota

ELEMENTARY STATISTICS
TWELFTH EDITION

Mario F. Triola
Dutchess Community College

PEARSON

Boston   Columbus   Indianapolis   New York   San Francisco   Upper Saddle River
Amsterdam   Cape Town   Dubai   London   Madrid   Milan   Munich   Paris   Montreal   Toronto
Delhi   Mexico City   São Paulo   Sydney   Hong Kong   Seoul   Singapore   Taipei   Tokyo

The author and publisher of this book have used their best efforts in preparing this book. These efforts include the development, research, and testing of the theories and programs to determine their effectiveness. The author and publisher make no warranty of any kind, expressed or implied, with regard to these programs or the documentation contained in this book. The author and publisher shall not be liable in any event for incidental or consequential damages in connection with, or arising out of, the furnishing, performance, or use of these programs.

Reproduced by Pearson from electronic files supplied by the author.

Copyright © 2014, 2010, 2007 Pearson Education, Inc.
Publishing as Pearson, 75 Arlington Street, Boston, MA 02116.

ISBN-13: 978-0-321-83799-8
ISBN-10: 0-321-83799-1

1 2 3 4 5 6 EBM 17 16 15 14 13

www.pearsonhighered.com

Contents

	Triola Elementary Statistics Page	Excel Manual Page

Getting Started with Microsoft® Excel

Overview

This manual is intended as a companion to Triola's *Elementary Statistics*, *12th ed.* It presents instructions on how to use Microsoft® Excel 2013 to carry out selected examples and exercises from *Elementary Statistics*, *12th ed.*

The first chapter of the manual contains an introduction to Microsoft ® Excel 2013 and how to perform basic operations such as entering data, sorting data, saving worksheets, and retrieving worksheets that have already been saved. All the screens pictured in this manual were obtained using Microsoft® Excel 2013 Preview on a PC with Windows 7. You may notice slight differences if you are using a different version of Excel or a different computer.

GS-1 Getting Started with the User Interface

GS-1.1	Excel 2013 Window

The figure shown below presents the top left side of the Excel 2013 window. Near the top of the figure, you see a row of several tabs: FILE, HOME, INSERT, PAGE LAYOUT, FORMULAS, DATA, REVIEW, VIEW, and ADD-INS. The HOME ribbon, shown below, presents groups of related commands. The groups displayed in the figure are Clipboard, Font, Alignment, and Number.

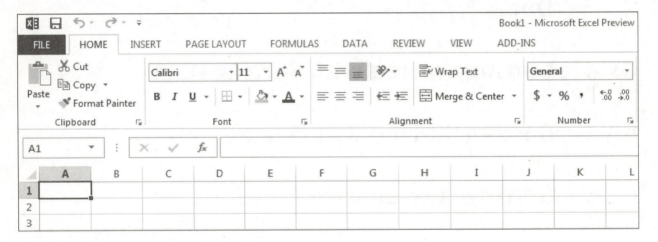

GS-1.2	FILE Tab and Ribbons

When you click the **FILE** tab, you will see a menu of commands along the left side of the screen that allow you to carry out operations on an Excel file, such as open a new file, retrieve a previously saved file, and so on. Some of these commands are New, Open, Save, Save As, and Print. When you click on the other tabs, you will see a ribbon of related commands. You will use some of the commands much more than others.

HOME ribbon: Clipboard, Font, Alignment, Number, Styles, Cells, and Editing.

INSERT ribbon: Tables, Illustrations, Apps, Charts, Reports, Sparklines, Filter, Links, Text, and Symbols.

PAGE LAYOUT ribbon: Themes, Page Setup, Scale to Fit, Sheet Options, and Arrange.

FORMULAS ribbon: Function Library, Defined Names, Formula Auditing, and Calculation.

DATA ribbon: Get External Data, Connections, Sort & Filter, Data Tools, Outline, and Analysis.

REVIEW ribbon: Proofing, Language, Comments, and Changes.

VIEW ribbon: Workbook View, Show, Zoom, Window, and Macros.

ADD-INS ribbon. This ribbon will contain any menu items that have been added through VBA macros or add-ins.

GS-1.3	# Dialog Boxes

Many of the statistical analysis procedures presented in this manual are associated with commands that are followed by dialog boxes. The dialog boxes usually require you to enter your choices or to select from alternatives that are presented. For example, many of the procedures explained in this manual are carried out by one of Excel's functions. If you click the **FORMULAS** tab and select **Insert Function**, a dialog box like the one shown below will appear. You make your category and

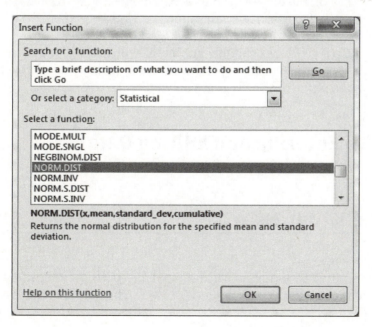

function selections by clicking on them. The **NORM.DIST** function selected in this dialog box was found in the **Statistical** category. If you don't know the name of a function or would like to find out what functions are available, you can type in a brief description of what you would like to do and then click the Go button. When you click the OK button at the bottom of the dialog box, another dialog box will often be displayed that asks you to provide information regarding the location of the data in the Excel worksheet.

GS-2 Getting Started with Opening Files

GS-2.1	**Opening a New Workbook**

When you start Excel, the screen opens to **Sheet 1** of **Book 1**. Sheet names appear on the tabs at the bottom of the screen. The name "Book 1" will appear at the top.

If you are already working in Excel and have finished one analysis problem and would like to open a new book for another problem, follow these steps. First, click the **FILE** tab at the top of the screen. Select **New**. Double-click on **Blank workbook**. If you were previously working in Book 1, the new file will be given the default name Book 2.

GS-2.2	**Opening a File That Has Already Been Created**

To open a file that you or someone else has already created, click on the **FILE** tab and select **Open**. A list of file locations will appear. Select the location by clicking on it. Many of the data files that are presented in your statistics textbook are available on the CD that accompanies the textbook. To open any of these files, you will select the CD drive on your computer.

After you select the CD drive, a list of folders and files available on the CD will appear. You select the folder or file you want by clicking on it. If you select a folder, another screen will appear with a list of files contained in the folder. Click on the name of the file that you would like to open.

GS-3 Getting Started with Entering and Editing Data

GS-3.1	**Cell Addresses**

Columns of the Excel worksheet are identified by letters of the alphabet and rows are identified by numbers. The cell address B1 refers to the cell located in column B row 1. The dark outline around a cell means that it is "active" and is ready to receive data. In the figure, cell B1 is ready to receive data. You will also see B1 in the Name Box to the left of the Formula Bar.

You can also activate a **range** of cells in an Excel worksheet. To activate a range of cells, first click in the top cell and drag down and across (or click in the bottom cell and drag up and across). The range of cells highlighted in the figure below is designated **C2:E4**.

GS-3.2	**Entering Data**

To enter data into a cell of an Excel worksheet, first activate the cell by clicking it. Then key in the desired data and press [**Enter**]. Pressing the [Enter] key moves you down to the next cell in that column. To enter the data in the figure shown below, you would follow these steps:

1. Click in cell **B1**. Key in **Females**. Press [**Enter**].
2. Key in **5**. Press [**Enter**].
3. Key in **2**. Press [**Enter**].

	A	B	C
1		Females	
2		5	
3		2	
4			
5			

GS-3.3	**Copying Data**

Excel provides a couple of different methods to copy data. One method will be illustrated here. To copy the data in one cell to another cell, follow these steps:

1. Click in the source cell.
2. Right-click and select **Copy** from the menu that appears.
3. Click in the target cell where you want the data to be placed.
4. Right-click and select one of the **Paste** options. Generally, the leftmost option in the top row will provide the result that you want.

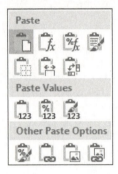

To copy of a **range** of cells to another location in an Excel worksheet, follow these steps:

1. Click and drag over the range of cells that you want to copy so that they are highlighted.
2. Right-click and select **Copy**.
3. Click in the topmost cell of the target location. Right-click and select one of the **Paste** options. Again, the leftmost option in the top row generally provides the result that you want.

GS-3.4	**Moving Data**

If you would like to move the contents of one cell from one location to another in an Excel worksheet, follow these steps:

1. Click in the cell containing the data that you would like to move.
2. Right-click and select **Cut** from the menu that appears.
3. Click in the target cell where you want the data to be placed.
4. Right-click and select **Paste**.

If you make a mistake, just click the Undo arrow located in the upper-left corner of the screen. It looks like this: ↶

GS-3.5	**Changing the Column Width**

There are a couple of different ways you can use to change the column width. Only one way will be described here. The output from Excel's Descriptive Statistics Data Analysis Tool will be used as an example. As you can see in the output displayed in the figure below, some of the labels in column A can only be partially viewed because the default column width is too narrow.

	A	B	C
1	*AGE*		
2			
3	Mean	8.438017	
4	Standard I	0.323738	
5	Median	8	
6	Mode	6	
7	Standard I	3.561115	

To increase the width of column A, follow these steps:

1. Position the mouse pointer directly on the vertical line between A and B in the letter row at the top of the worksheet so that it turns into a black plus sign.
2. Click and drag to the right until you can read all the output labels. (You can also click and drag to the left to make columns narrower.) After increasing the width of column A, your output should appear similar to the output shown below.

	A	B	C
1	*AGE*		
2			
3	Mean	8.438016529	
4	Standard Error	0.323737761	
5	Median	8	
6	Mode	6	
7	Standard Deviation	3.561115373	

GS-4 Getting Started with Sorting Data

GS-4.1	## Sorting a Single Column of Data

Let's say that you have entered the ages of six research participants in column A of an Excel worksheet and that you would like to sort the age values in ascending order.

	A
1	AGE
2	10
3	9
4	11
5	6
6	8
7	6

1. Click and drag from cell A1 to cell A7 so that the range **A1:A7** is highlighted.

You could also click directly on A in the letter row at the top of the worksheet. This results in all cells of column A being highlighted.

---

2. At the top of the screen, click the **DATA** tab.
3. Click on the word **Sort** in the Sort & Filter group.
4. In the Sort dialog box, select to sort by **AGE**, to sort on **Values**, and to sort in order of **Smallest to Largest**.

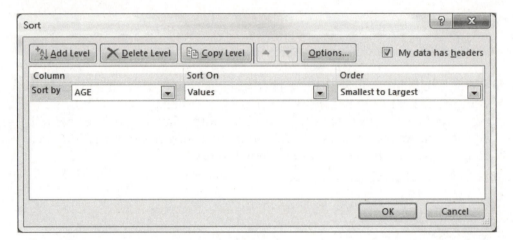

5. Click **OK**. The data in column A should now be assorted in ascending order.

	A
1	AGE
2	6
3	6
4	8
5	9
6	10
7	11

GS-4.2	**Sorting Multiple Columns of Data**

Your Excel data files will often contain multiple columns of data. Let's say that you have an Excel data file that contains the data shown at the top of the next page and that you like to sort the file by ABSENCES in descending order.

	A	B	C
1	AGE	IQ	ABSENCES
2	10	97	0
3	9	77	5
4	11	128	1
5	6	99	1
6	8	80	3
7	6	118	0

1. Click and drag from cell A1 to cell C7 so that the range **A1:C7** is highlighted.
2. At the top of the screen, click the **DATA** tab.
3. Click on the word **Sort** in the Sort & Filter group.
4. In the Sort dialog box that appears, you are given the option of sorting the data by three different variables. You want to sort by ABSENCES in descending order. Click the arrow to the right of the Sort by window.
5. Click on **ABSENCES** to select it.

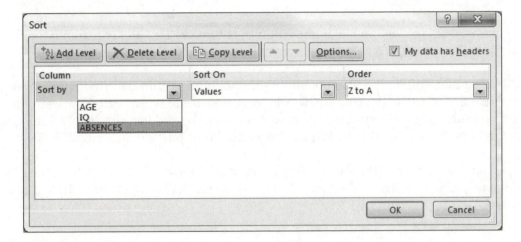

6. Select to sort on **Values** and select to sort in the order of **Z to A** (largest to smallest). Click **OK**.

The sorted data file is shown below.

	A	B	C
1	AGE	IQ	ABSENCES
2	9	77	5
3	8	80	3
4	11	128	1
5	6	99	1
6	10	97	0
7	6	118	0

GS-5 Getting Started with Saving Data

GS-5.1	**Saving Files**

To save a newly created file for the first time, click on the **FILE** tab in the upper-left corner of the screen and select **Save**. A list of locations will appear. You select the location for saving the file by double-clicking on it.

GS-5.2	**Naming Files**

The default file name, displayed in the File name window, is **Book1**. It is highly recommended that you replace the default name with a name that is more descriptive of the file contents.

Windows and Mac versions of Excel will allow files names to have around 200 characters. You will find that descriptive file names with complete words will be easier to work with than really short names with abbreviations. For example, if a file contains data collected in a survey of Minneapolis residents, you might want to name the file **Minneapolis Resident Survey** rather than **MRS**.

Several symbols cannot be used in file names. These include the forward slash (/), backslash (\), greater-than sign (>), less-than sign (<), asterisk (*), question mark (?), quotation mark ("), pipe symbol (|), colon (:), and semicolon (;).

GS-6 Getting Started with Add-Ins

GS-6.1	**Loading Excel's Analysis ToolPak**

The Analysis ToolPak is an Excel Add-in that you will probably use a great deal for statistical analyses. If Data Analysis does not appear in the Analysis group of the DATA ribbon as shown below, then you will need to load it.

1. First click the **FILE** tab and select **Options**.

2. Select **Add-Ins** from the list on the left.
3. Analysis ToolPak and Analysis ToolPak-VBA should both be in the list of Active Application Add-ins. If you need to add one or both of them, first click on the name to select it. Then click **Go** at the bottom of the dialog box.

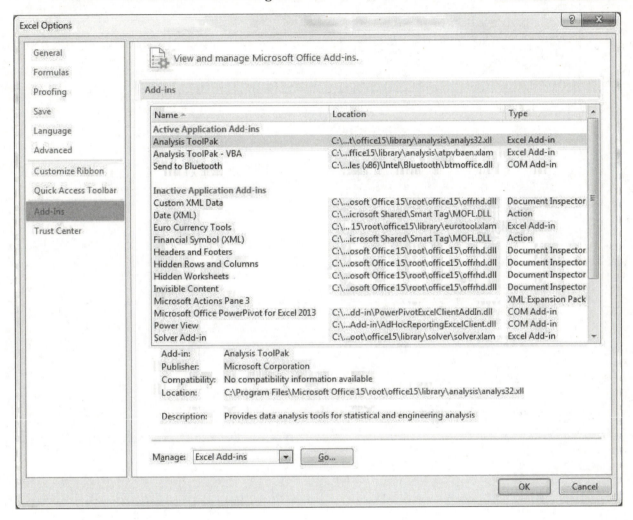

4. In the Add-Ins dialog box, place a check mark in the box next to the add-in to make it active. Click **OK**.

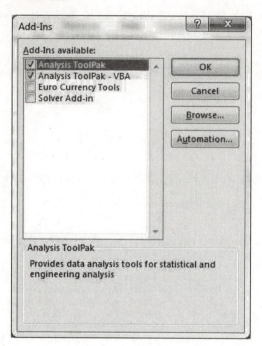

2

Summarizing and Graphing Data

2-2 Frequency Distributions

Example 1, page 46	Using XLSTAT to Construct a Frequency Distribution and a Relative Frequency Distribution of IQ Scores

Example 1 asks you to construct a frequency distribution like the one displayed in Table 2-2 on page 45 of the Triola text. You will be using the IQ scores of the low lead group in Table 2-1 on page 43 of the text. These data are also available in the IQLEAD file on your data disk. The output from XLSTAT's histogram procedure includes both a frequency distribution and a relative frequency distribution.

1. Open the **IQLEAD** file on your data disk. The first few rows are shown below. The first column, LEAD, indicates the lead level group. You will be constructing a frequency distribution of full-scale IQ for group 1, the low lead group. The last column, IQF, is full-scale IQ.

	A	B	C	D	E	F	G	H
1	LEAD	AGE	SEX	YEAR1	YEAR2	IQV	IQP	IQF
2	1	11	1	25	18	61	85	70
3	1	9	1	31	28	82	90	85
4	1	11	1	30	29	70	107	86

2. At the top of the screen, select the **XLSTAT** add-in.
3. Select **Describing data**. Select **Histograms**.
4. Complete the General dialog box as shown below. A description of the entries immediately follows the dialog box.

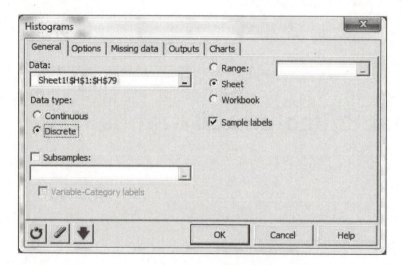

- **Data**: Enter the range of the IQF data for group 1, the low lead group, **H1:H79**.
- **Data type**: Select **Discrete**.
- **Sheet**: The output will be placed in a new Excel worksheet.

- **Sample labels**: Because the top cell of the data range is a label, you need to select **Sample labels**.

5. Click the **Options** tab. The instructions in the text tell you to use a class width of 20. **Range** refers to the class width. Select **Range** and enter **20**.
6. Click the **Outputs** tab. Select **Descriptive statistics**.
7. Click the **Charts** tab. Select **Histograms** and **Bars** as shown below. For ordinate of the histograms, select **Frequency**.

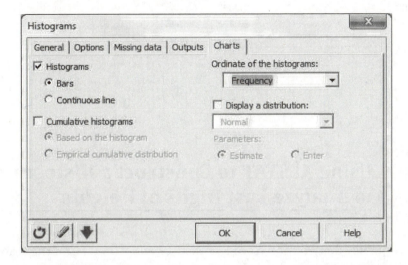

8. Click **OK**. Click **Continue** in the XLSTAT-Selections dialog box.

The XLSTAT output includes descriptive statistics for the IQF variable, a histogram, and descriptive statistics for the intervals. The XLSTAT histogram output is shown below.

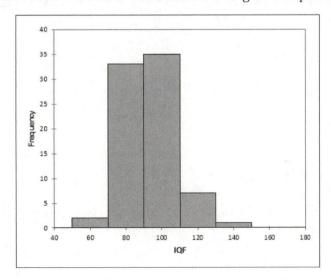

The descriptive statistics for the intervals provided by XLSTAT (shown below) include upper and lower bounds, frequencies, relative frequencies, and densities.

Descriptive statistics for the intervals :				
Lower bound	Upper bound	Frequency	Relative frequency	Density
50	70	2	0.0256	0.0013
70	90	33	0.4231	0.0212
90	110	35	0.4487	0.0224
110	130	7	0.0897	0.0045
130	150	1	0.0128	0.0006
150	170	0	0.0000	0.0000

2-3   Histograms

Exercise 9, page 59	Using XLSTAT to Construct a Histogram to Analyze Last Digits of Heights

Exercise 9 asks you to use the frequency distribution from Exercise 19 in Section 2-2 of the Triola text to construct a histogram. You will first enter the data in an Excel worksheet and then use XLSTAT to construct a histogram.

1. Enter the data from Exercise 19 in Section 2-2 on page 52 of the Triola text in an Excel worksheet. The first few rows are shown below. Be sure to enter the label, **Last Digit**, in cell **A1**.

	A
1	Last Digit
2	0
3	0
4	0
5	0

2. At the top of the screen, select the **XLSTAT** add-in.
3. Select **Visualizing data**. Select **Histograms**.
4. Complete the General dialog box as shown at the top of the next page. A description of the entries immediately follows the dialog box.

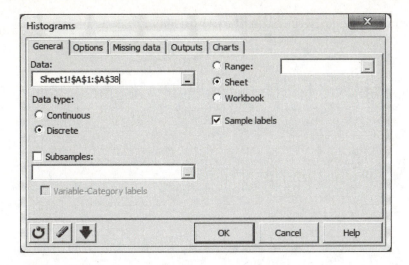

- **Data**: Enter the range of the Last Digit data, **A1:A38**.
- **Data type**: Select **Discrete**.
- **Sheet**: The output will be placed in a new Excel worksheet.
- **Sample labels**: Because the top cell of the data range is a label, you need to select **Sample labels**.

5. Click the **Options** tab. **Number** refers to the number of classes you would like. Exercise 19 in Section 2-2 indicates that the frequency distribution should have 10 classes. Select **Number** and enter **10** for the number of classes.

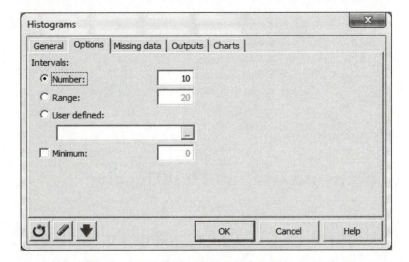

6. Click the **Charts** tab. Select **Histograms** and **Bars**. For ordinate of the histograms, select **Frequency**.
7. Click **OK**. Click **Continue** in the XLSTAT-Selections dialog box.

The histogram and descriptive statistics provided in the XLSTAT output are shown below.

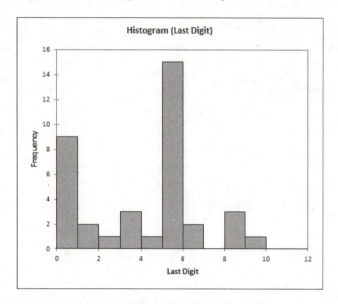

Descriptive statistics for the intervals :				
Lower bound	Upper bound	Frequency	Relative frequency	Density
0	1	9	0.2432	0.2432
1	2	2	0.0541	0.0541
2	3	1	0.0270	0.0270
3	4	3	0.0811	0.0811
4	5	1	0.0270	0.0270
5	6	15	0.4054	0.4054
6	7	2	0.0541	0.0541
7	8	0	0.0000	0.0000
8	9	3	0.0811	0.0811
9	10	1	0.0270	0.0270

2-4   Graphs That Enlighten and Graphs That Deceive

Example 1, page 61

Using XLSTAT to Construct a Scatterplot of Waist and Arm Circumferences

Example 1 presents an interpretation of the scatterplot in Figure 2-6 of paired waist/arm measurements of randomly selected males. These data are in the MBODY data file on your data disk. You will be using XLSTAT to construct this scatterplot.

1.   Open the **MBODY** data file on your data disk. There are 14 variables in the data set. The first few rows of columns J through N are shown at the top of the next page. WAIST is the

variable label for waist circumference (cm) (column L), and ARMC is the variable label for arm circumference (cm) (column M).

	J	K	L	M	N
WT		**HT**	**WAIST**	**ARMC**	**BMI**
	64.4	178.8	81.4	28.4	20.14
	61.8	177.5	74.8	26.8	19.62
	78.5	187.8	84.1	32.3	22.26
	86.3	172.4	95.5	39.0	29.04

2. At the top of the screen, select the **XLSTAT** add-in.
3. Select **Visualizing data**. Select **Scatter plots**.
4. Complete the General dialog box as shown below. A description of the entries immediately follows the dialog box.

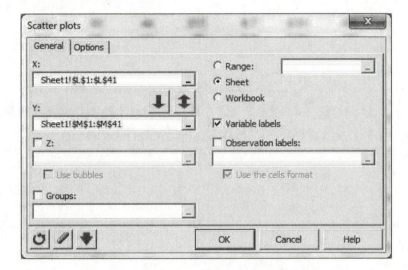

- **X**: The *X* variable for the scatterplot is WAIST. Enter the data range for WAIST, **L1:L41**.
- **Y**: The *Y* variable for the scatterplot is ARMC. Enter the data range for ARMC, **M1:M41**.
- **Sheet**: The output will be placed in a new Excel worksheet.
- **Variable labels**: Because the top cells in the data ranges are labels, you need to select **Variable labels**.

5. Click **OK**. Click **Continue** in the XLSTAT-Selections dialog box.

The scatterplot provided in the XLSTAT output is shown below.

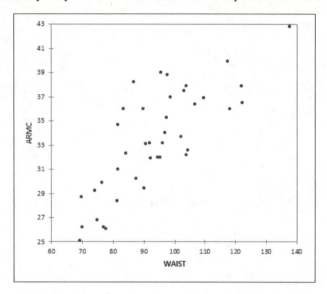

<table>
<tr><td rowspan="2" style="background:#d9d9d9;">Example 6, page 63</td><td>Using XLSTAT to Construct a Stemplot of</td></tr>
<tr><td>IQ Scores of the Low Lead Group</td></tr>
</table>

Example 6 presents a stemplot that displays the IQ scores of the low lead group. These IQ scores are shown in Table 2-1 on page 43 of the Triola text. These data are also available in the IQLEAD file on your data disk. You will use XLSTAT to construct a stemplot similar to the one shown in Example 6.

1. Open the **IQLEAD** file on your data disk. The first few rows are displayed below. The first column, LEAD, indicates the lead level group. You will be constructing a stemplot of full-scale IQ for group 1, the low lead group. The last column, IQF, is full-scale IQ.

	A	B	C	D	E	F	G	H
1	LEAD	AGE	SEX	YEAR1	YEAR2	IQV	IQP	IQF
2	1	11	1	25	18	61	85	70
3	1	9	1	31	28	82	90	85
4	1	11	1	30	29	70	107	86

2. At the top of the screen, select the **XLSTAT** add-in.
3. Select **Visualizing data**. Select **Univariate plots**.
4. Complete the General dialog box as shown at the top of the next page. A description of the entries immediately follows the dialog box.

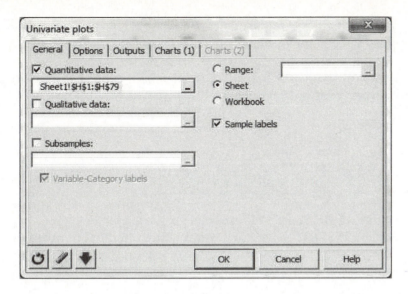

- **Data**: Enter the range of the IQF data for the low lead group, **H1:H79**.
- **Sheet**: The output will be placed in a new Excel worksheet.
- **Sample labels**: Because the top cell of the data range is a label, you need to select **Sample labels**.

5. Click the **Options** tab. Select **Charts**.
6. Click the **Charts(1)** tab. Select **Stem-and-leaf plots**. Click **OK**. Click **Continue** in the XLSTAT-Selections dialog box.

The XLSTAT stem-and-leaf plot of full-scale IQ for the low lead group is shown below.

Stem-and-leaf plot (IQF):		
Unit:	10	
5	0 6	
6		
7	0 2 3 4 5 6 6 6 6 6 7 7 8	
8	0 0 0 4 5 5 5 5 6 6 6 6 7 7 8 8 8 9 9 9	
9	1 2 3 4 4 4 5 6 6 6 6 6 6 6 7 7 8 9 9 9 9	
10	0 1 1 2 4 4 5 5 6 7 7 7 7 8	
11	1 5 5 8	
12	0 5 8	
13		
14	1	

Example 7, page 64	**Using Excel's Charts to Construct a Multiple Bar Graph of Income by Gender**

Example 7 presents a multiple bar graph of the median incomes of males and females for different years. You will be constructing a similar bar graph using data from the U. S. Census (http://www.census.gov/hhes/www/income/data/historical/people/). The data were taken from a summary table that displayed median income by gender for the years 1953 to 2011.

1. Enter the year, gender, and median income data in an Excel worksheet as shown below. Median income is displayed in thousands of dollars.

	A	B	C	D	E	F	G
1		1960	1970	1980	1990	2000	2010
2	Male	4.08	6.67	12.53	20.293	28.343	32.205
3	Female	1.261	2.237	4.92	10.07	16.079	20.775

2. Click and drag over the range **A1:G3** to select these cells for the graph.
3. At the top of the screen, select **INSERT**. Select the **Column chart** in the top row of the Charts group.
4. Select the leftmost chart in the 2-D Column row.

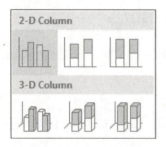

5. Click directly on **Chart Title** to change the title to **Median Income by Gender**.
6. To add other chart elements, click anywhere within the chart and then click on the plus sign. Select **Axis Titles**.
7. Click on the *y*-axis **Axis Title** and type **Median Income (thousands of dollars)**. Press [**Enter**].
8. Click on the *x*-axis **Axis Title** and type **Year**. Press [**Enter**].

The Excel column chart is shown below.

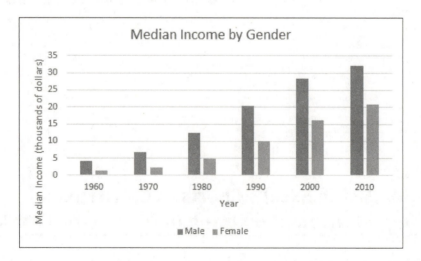

| Example 8, page 64 | Using Excel's Charts to Construct a Pareto Chart |

Example 8 presents a pareto chart where response categories are arranged from highest to lowest frequency. You will construct a pareto chart for the ten states in the United States that have the highest percent of people with a bachelor's degree or more. The data were taken from a U.S. Census summary table that displayed educational level by state as of 2009 (http://www.census.gov/compendia/statab/cats/education/educational_attainment.html).

1.   Enter the state and percent data in columns B and C as shown below. You do not need to enter the state names that appear in column A. These names are provided as a reference in case you are not certain of all the states' abbreviations. Note that the data are already sorted by highest to lowest percent.

	A	B	C
1		State	Percent
2	Massachusetts	MA	38.2
3	Colorado	CO	35.9
4	Maryland	MD	35.7
5	Connecticut	CT	35.6
6	New Jersey	NJ	34.5
7	Virginia	VA	34.0
8	Vermont	VT	33.1
9	New York	NY	32.4
10	New Hampshire	NH	32.0
11	Minnesota	MN	31.5

2.   Click and drag over the range **B1:C11** to select the state and percent data for the graph.
3.   At the top of the screen, select **INSERT**. Select the **Column chart** in the top row of the Charts group.
4.   Select the leftmost chart in the 2-D Column row.

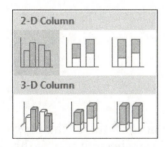

5.   To modify or add chart elements, click anywhere within the chart and then click on the plus sign. Select **Axis Titles**.
6.   Click on the *y*-axis **Axis Title** and type **Percent**. Press [**Enter**].
7.   Click on the *x*-axis **Axis Title** and type **State**. Press [**Enter**].
5.   Click on **Percent** (the chart title) and type **Percent with Bachelor's Degree or More by State**. Press [**Enter**].

The Excel column chart is shown below.

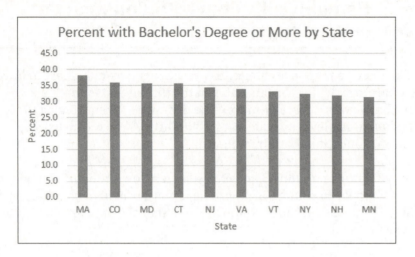

Example 9, page 65	**Using Excel's Charts to Construct a Pie Chart**

Example 9 in the Triola text presents a pie chart that displays data related to what contributes most to people's happiness. To learn how to use Excel to construct a pie chart, you will use data regarding how people spend their time. The data come from the Bureau of Labor Statistics, 2011 American Time Use Survey (http://www.bls.gov/tus/charts/). The pie chart you will construct will display time use on an average work day for employed persons ages 25 to 54 with children.

1. Enter the activity and hours data in an Excel worksheet as shown below.

	A	B
1	Activity	Hours
2	Sleeping	7.6
3	Working and related activities	8.8
4	Leisure and sports	2.5
5	Household activities	1.1
6	Eating and drinking	1.1
7	Caring for others	1.2
8	Other	1.7

2. Click and drag over the range **A1:B8** to select the activity and hours data for the pie chart.
3. At the top of the screen, select **INSERT**. Select the **Pie Chart** in the bottom row of the Charts group.
4. Select the leftmost chart in the 2-D Pie row as shown at the top of the next page.

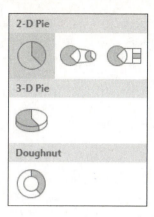

5. Because we want to add labels for the hours and activities, the chart needs to be larger. Click anywhere within the figure so that handles appear on the edges. Click and drag the handles to make the figure taller and wider.

6. To modify or add chart elements, click anywhere within the chart and then click on the plus sign. Select **Chart Title** and **Data Labels.** For data labels, select **Data Callout**.

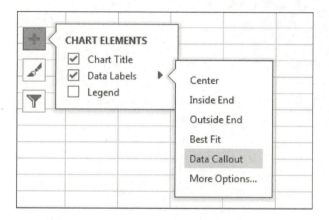

7. Click directly on the chart title, **Hours**, and type **Time Use on an Average Work Day**. Press [**Enter**].

The Excel pie chart is shown below. If you would like, you can display percentages in the data labels. To make this change or others, select Data Labels ⇾ More Options.

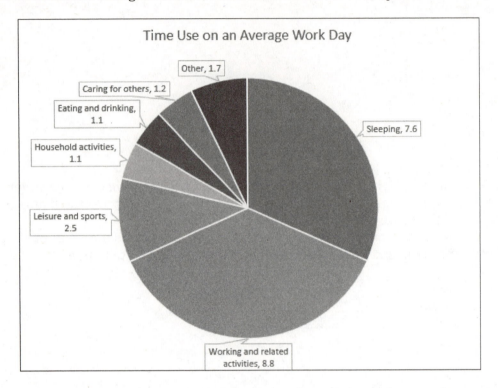

3

Statistics for Describing, Exploring, and Comparing Data

3-2 Measures of Center

Examples 1, 2, 3, and 4, pages 82-84	Using Data Analysis Tools to Find the Mean, Median, and Mode of the Number of Chocolate Chips

We will first use Excel's Data Analysis Tools to find the mean, median, and mode of the number of chocolate chips in Chips Ahoy (regular) displayed in Table 3-1 on page 79 of the Triola text. Then we will use XLSTAT to find the mean and median. XLSTAT does not report the mode for quantitative variables.

1. Enter the Chips Ahoy (regular) data in column A of an Excel worksheet. The first few rows are shown below.

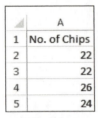

	A
1	No. of Chips
2	22
3	22
4	26
5	24

2. Click **DATA** at the top of the screen and select **Data Analysis** on the far right of the data ribbon.

If Data Analysis does not appear as a choice in the data ribbon, you will need to load the Microsoft Excel ToolPak add-in. Follow the procedure on page 10.

3. Select **Descriptive Statistics**. Click **OK**.

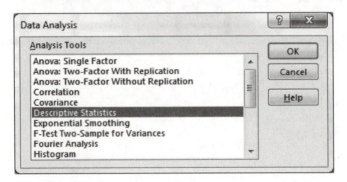

4. Complete the dialog box as shown at the top of the next page. A description of the entries immediately follows the dialog box.

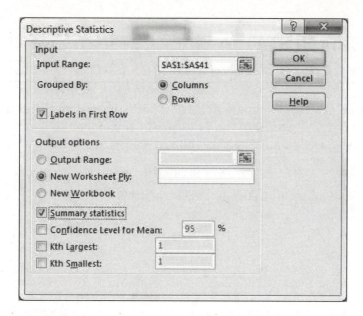

- **Input Range**: Enter the range of the chips data, **A1:A41**.
- **Grouped By**: Because the data appear in a column of the worksheet, you need to select **Columns**.
- **Labels in First Row**: Select **Labels in First Row** to let Excel know that the top cell in the data range is a label and not a data value.
- **Output options**: Select **New Worksheet Ply** to place the output in a new Excel worksheet.
- **Summary statistics**: Be sure to select **Summary statistics**, otherwise you will not get any output.

5. Click **OK**.

The Excel descriptive statistics output is displayed below. The mean is 23.95, the median is 24, and the mode is 23. Note that the Excel output will only present one mode, even if the distribution is multimodal. Therefore, if you are not familiar with the data, it's a good idea to construct a frequency distribution before you obtain the descriptive statistics.

	A	B
1	*No. of Chips*	
2		
3	Mean	23.95
4	Standard Error	0.403431
5	Median	24
6	Mode	23
7	Standard Deviation	2.55152
8	Sample Variance	6.510256
9	Kurtosis	-0.13805
10	Skewness	-0.0876
11	Range	11
12	Minimum	19
13	Maximum	30
14	Sum	958
15	Count	40

| Examples 1, 2, 3, and 4, pages 82-84 | Using XLSTAT to Find the Mean and Median of the Number of Chocolate Chips |

This time, we will use XLSTAT to find the mean and median of the number of chocolate chips in Chips Ahoy (regular) displayed in Table 3-1 on page 79 of the Triola text. If you would like the mode to be included, you will need to use Excel's Data Analysis Tools, because XLSTAT does not report the mode for quantitative variables.

1. Enter the Chips Ahoy (regular) data in column A of an Excel worksheet. The first few rows are shown below.

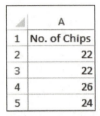

2. At the top of the screen, select the **XLSTAT** add-in.
3. Select **Describing data**. Select **Descriptive statistics**.
4. Complete the General dialog box as shown below. A description of the entries is given immediately after the dialog box.

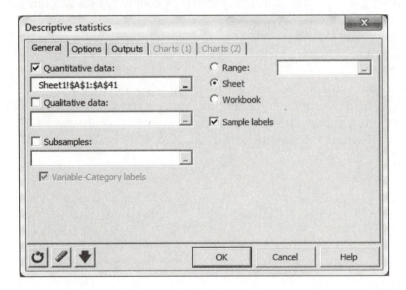

- **Quantitative data**: Enter the data range, **A1:A41.**
- **Sheet**: The output will be placed in a new Excel worksheet.
- **Sample labels**: Because the top cell in the data range is a label and not a data value, you need to select **Sample labels**.

5. Click the **Options** tab. Select **Descriptive statistics**.
6. Click the **Outputs** tab. XLSTAT allows you to select the descriptive statistics that you would like included in the output. For this example, let's use the default selection.

7. Click **OK**. Click **Continue** in the XLSTAT-Selections dialog box.

The XLSTAT output is displayed below. The median is 24 and the mean is 23.95. Recall that XLSTAT does provide the mode as a summary measure of quantitative data.

Descriptive statistics (Quantitative data):	
Statistic	No. of Chips
No. of observations	40
Minimum	19.0000
Maximum	30.0000
1st Quartile	22.7500
Median	24.0000
3rd Quartile	26.0000
Mean	23.9500
Variance (n-1)	6.5103
Standard deviation (n-1)	2.5515

3-3 Measures of Variation

Example 8, page 105	**Using Data Analysis Tools to Find the Range, Standard Deviation, and Variance of Weights of Coke**

We will first use Excel's Data Analysis Tools to find the range, standard deviation, and variance of weights of Coke. Then, we will use XLSTAT. The data are located in the COLA file on your data disk.

1. Open the **COLA** file on your data disk. The first few rows of the data set shown below. We will be obtaining descriptive statistics for weight of regular coke in column A. The abbreviated label is CKREGWT.

	A	B	C	D	E	F	G	H
1	CKREGWT	CKREGVOL	CKDIETWT	CKDTVOL	PPREGWT	PPREGVOL	PPDIETWT	PPDTVOL
2	0.8192	12.3	0.7773	12.1	0.8258	12.4	0.7925	12.3
3	0.815	12.1	0.7758	12.1	0.8156	12.2	0.7868	12.2
4	0.8163	12.2	0.7896	12.3	0.8211	12.2	0.7846	12.2
5	0.8211	12.3	0.7868	12.3	0.817	12.2	0.7938	12.3

2. Click **DATA** at the top of the screen and select **Data Analysis** on the far right of the data ribbon.

If Data Analysis does not appear as a choice in the data ribbon, you will need to load the Microsoft Excel ToolPak add-in. Follow the procedure on page 10.

3. Select **Descriptive Statistics**. Click **OK**.

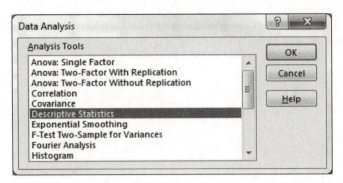

4. Complete the dialog box as shown below. Be sure to select **Summary statistics**, otherwise you will not get any output.

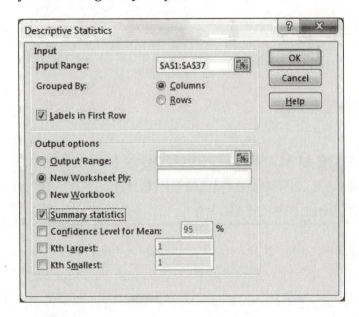

5. Click **OK**.

The Excel descriptive statistics output is displayed below. The range is 0.0394, the standard deviation is 0.00751, and the variance is 5.64E-05. Both the standard deviation and the variance were calculated using the sample formulas.

	A	B
1	*CKREGWT*	
2		
3	Mean	0.816822
4	Standard Error	0.001251
5	Median	0.8171
6	Mode	0.8192
7	Standard Deviation	0.007507
8	Sample Variance	5.64E-05
9	Kurtosis	3.476007
10	Skewness	-1.23818
11	Range	0.0394
12	Minimum	0.7901
13	Maximum	0.8295
14	Sum	29.4056
15	Count	36

Example 8, page 105	# Using XLSTAT to Find the Range, Standard Deviation, and Variance of Weights of Coke

We will now use XLSTAT to obtain measures of variation for weights of regular Coke.

1. If you have not already done so, open the **COLA** file on your data disk. The weights of regular coke are in column A of the worksheet. The abbreviated label is CKREGWT.
2. At the top of the screen, select the **XLSTAT** add-in.
3. Select **Describing data**. Select **Descriptive statistics**.
4. Complete the General dialog box as shown at the top of the next page. A description of the entries immediately follows the dialog box.

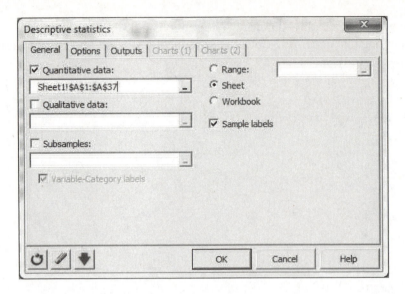

- **Quantitative data**: Enter the range of the CKREGWT data, **A1:A37**.
- **Sheet**: The output will be placed in a new Excel worksheet.
- **Sample labels**: Because the top cell in the data range is a label and not a data value, you need to select **Sample labels**.

5. Click the **Options** tab. Select **Descriptive statistics**.
6. Click the **Outputs** tab. XLSTAT allows you to select the descriptive statistics you would like included in the output. You can select the variance and standard deviation calculated using the population formulas or the sample formulas. The population formulas are designated Variance(n) and Standard deviation (n). The sample formulas are designated Variance (n-1) and Standard deviation (n-1). For this example, let's select **No. of observations**, **Minimum**, **Maximum**, **Range**, **Variance (n-1)**, and **Standard deviation (n-1)**.
7. Click **OK**. Click **Continue** in the XLSTAT-Selections dialog box.

The XLSTAT descriptive statistics output is displayed below on the left. By default, values with decimals are rounded to four decimal places. To display more decimal places, select the cell(s), **right-click** and select **Format Cells**. Click the **Number** tab. Select the **Number** category. Increase decimal places to the desired number of places. I chose **6** decimal places for the bottom five measures displayed in the XLSTAT descriptive statistics output on the right.

Descriptive statistics (Quantitative data):	
Statistic	CKREGWT
No. of observations	36
Minimum	0.7901
Maximum	0.8295
Range	0.0394
Variance (n-1)	0.0001
Standard deviation (n-1)	0.0075

Descriptive statistics (Quantitative data):	
Statistic	CKREGWT
No. of observations	36
Minimum	0.790100
Maximum	0.829500
Range	0.039400
Variance (n-1)	0.000056
Standard deviation (n-1)	0.007507

3-4 Measures of Relative Standing and Boxplots

Example 2, page 114	**Using the STANDARDIZE Function to Find the z Score for a Pulse Rate**

Example 2 asks you if a pulse rate of 48 is unusual. To answer this question, you will use Excel's STANDARDIZE function to convert 48 to a z score, assuming the distribution of pulse rates has a mean of 67.3 beats per minute and a standard deviation of 10.3 beats per minute.

1. At the top of the screen, click **FORMULAS**. Then click **Insert Function**.
2. Select the **Statistical** category. Select the **STANDARDIZE** function. Click **OK**.

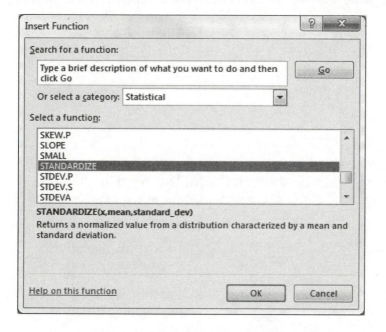

3. Complete the STANDARDIZE dialog box as shown below. The function returns -1.874.

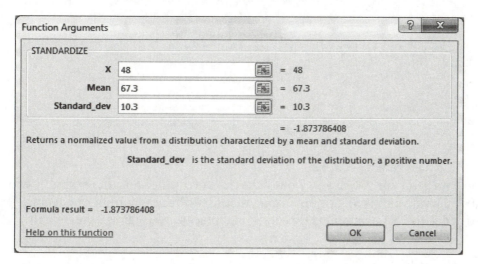

Example 3, page 115	**Using the PERCENTRANK.INC Function to Find the Percentile of a Data Value**

Example 3 asks you to find the percentile for a cookie with 23 chocolate chips in the Chips Ahoy regular distribution shown in Table 3-1. Your textbook states that there are a number of different procedures for calculating percentiles. Excel offers two functions for calculating percentiles, PERCENTRANK.INC and PERCENTRANK.EXC. For this example, we will use PERCENTRANK.INC. The results from PERCENTRANK.INC and PERCENTRANK.EXC will differ slightly based on the calculation procedure.

1. If you have not already done so, enter in an Excel worksheet the Chips Ahoy regular data displayed in Table 3-1 on page 79 of the Triola text. These data were used earlier in this manual in the first example in Section 3-2, Measures of Center. Click in cell **C1** where the result will be placed.

◢	A	B	C	D
1	No. of Chips			
2	22			
3	22			
4	26			
5	24			

2. At the top of the screen, click **FORMULAS**. Then click **Insert Function**.
3. Select the **Statistical** category. Select the **PERCENTRANK.INC** function. Click **OK**.
4. Complete the dialog box as shown below. A description of the entries is given immediately after the dialog box.

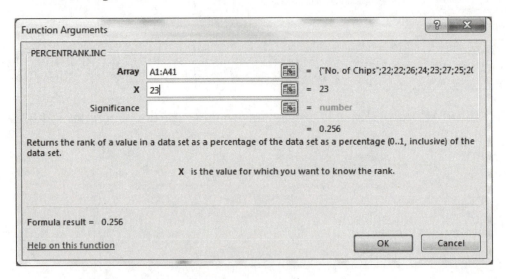

*   **Array**: Enter the data range, **A2:A41**.
*   **X**: Enter the data value for which you would like the percentile. For this example, enter **23**.
*   **Significance**: Significance is an optional value that identifies the number of decimal places for the returned value. The default is three decimal places. We will use the default.

5. Click **OK**. Excel's PERCENTRANK.INC function returns 0.256.

<table>
<tr><td>Example 4, page 115</td><td>**Using the PERCENTILE.INC Function to Convert a Percentile to a Data Value**</td></tr>
</table>

Example 4 asks you to convert the 18th percentile to a data value in the Chips Ahoy regular distribution shown in Table 3-1. Keep in mind that there are a number of different procedures for calculating percentiles. Excel provides two different functions for calculating percentiles, PERCENTILE.INC and PERCENTILE.EXC. We will use the PERCENTILE.INC function.

1. If you have not already done so, enter in an Excel worksheet the Chips Ahoy regular data displayed in Table 3-1 on page 79 of the Triola text. These data were used earlier in this manual in the first example in Section 3-2, Measures of Center. Click in cell **C1** where the output will be placed.

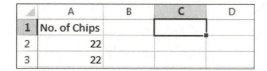

◢	A	B	C	D
1	No. of Chips			
2	22			
3	22			

2. At the top of the screen, click **FORMULAS**. Then click **Insert Function**.
3. Select the **Statistical** category. Select the **PERCENTILE.INC** function. Click **OK**.
4. Complete the dialog box as shown below. A description of the entries is given immediately after the dialog box.

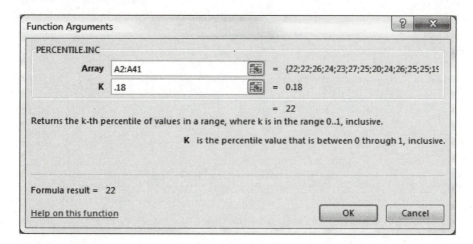

- **Array**: Enter the data range, **A2:A41**.
- **K:** Enter the percentile as a proportion. For this example, enter **.18** for the 18th percentile.

5. Click **OK**. Excel's PERCENTILE.INC function returns a data value of 22.

Example 6, page 117	**Using the QUARTILE.INC Function to Find a Quartile**

Example 4 asks you to find the value of the first quartile, Q_1, in the Chips Ahoy regular distribution. Excel provides two functions to find quartiles, QUARTILE.INC and QUARTILE.EXC. The two functions use slightly different calculation procedures. We will use the QUARTILE.INC function.

1. If you have not already done so, enter in an Excel worksheet the Chips Ahoy regular data displayed in Table 3-1 on page 79 of the Triola text. These data were used earlier in this manual in the first example in Section 3-2, Measures of Center. Click in cell **D1** where the output will be placed.

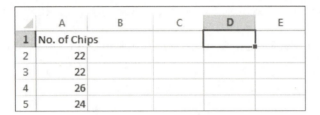

2. At the top of the screen, click **FORMULAS**. Then click **Insert Function**.
3. Select the **Statistical** category. Select **QUARTILE.INC**. Click **OK**.
4. Complete the dialog box as shown below. A description of the entries is given immediately after the dialog box.

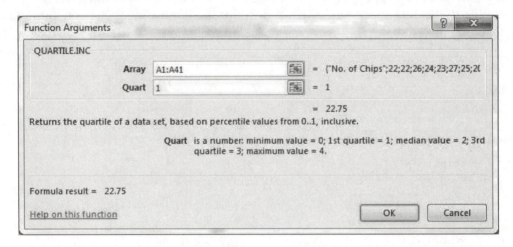

- **Array**: Enter the data range, **A1:A41**. Note that it does not matter whether or not the label in cell A1 is included in the range.
- **Quart**: The explanation in the dialog box tells you to enter 0 for the minimum value in the distribution, 1 for the 1st quartile, 2 for the median value (or 2nd quartile), 3 for the 3rd quartile, and 4 for the maximum value. Because we want the 1st quartile, we enter **1**.

5. Click **OK**. Excel's QUARTILE.INC function returns 22.75.

Example 7, page 118	# Using XLSTAT to Find a 5-Number Summary

Example 7 asks you to find the 5-number summary using the chocolate chip counts listed in Table 3-4. These are the same Chips Ahoy regular data that are presented Table 3-1, except that the data in Table 3-4 have been sorted in ascending order.

1. If you have not already done so, enter in an Excel worksheet the Chips Ahoy regular data displayed in Table 3-1 on page 79 of the Triola text. These data were used earlier in this manul in the first example in Section 3-2, Measures of Center. The first few lines are shown below.

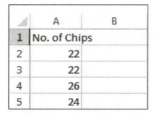

	A	B
1	No. of Chips	
2	22	
3	22	
4	26	
5	24	

2. At the top of the screen, select the **XLSTAT** add-in.
3. Select **Describing data**. Select **Descriptive statistics**.
4. Complete the General dialog box as shown below. A description of the entries is shown immediately after the dialog box.

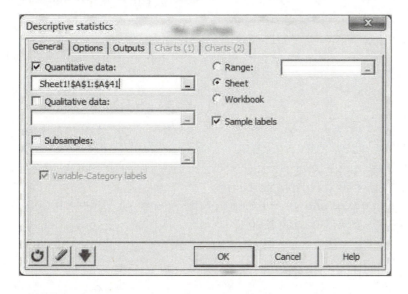

*   **Quantitative data**: Enter the data range, **A1:A41**.
*   **Sheet**: The output will be placed in a new Excel worksheet.
*   **Sample labels**: Because the top cell in the data range is a label, you need to select **Sample labels**.

5. Click the **Outputs** tab. For the 5-number summary, select **Minimum**, **Maximum**, **1st Quartile**, **Median**, and **3rd Quartile**.
6. Click **OK**. Click **Continue** in the XLSTAT-Selection dialog box.

The XLSTAT descriptive statistics output is displayed below. The minimum is 19, the maximum is 30, the 1st quartile is 22.75, the median is 24, and the 3rd quartile is 26.

Descriptive statistics (Quantitative data):			
Statistic	No. of Chips		
Minimum	19.0000		
Maximum	30.0000		
1st Quartile	22.7500		
Median	24.0000		
3rd Quartile	26.0000		

Example 8, page 119   **Using XLSTAT to Construct a Boxplot**

Example 8 asks you to construct a boxplot for Chips Ahoy regular using the chocolate chip counts listed in Table 3-4. These are the same Chips Ahoy regular data presented Table 3-1, except that the data in Table 3-4 have been sorted in ascending order.

1. If you have not already done so, enter in an Excel worksheet the Chips Ahoy regular data displayed in Table 3-1 on page 79 of the Triola text. These data were used earlier in this manual in the first example in Section 3-2, Measures of Center. The first few lines are shown below.

	A	B
1	No. of Chips	
2	22	
3	22	
4	26	
5	24	

2. At the top of the screen, click the **XLSTAT** add-in.
3. Select **Describing data**. Select **Descriptive statistics**.
4. Complete the General dialog box as shown at the top of the next page. A description of the entries is shown immediately after the dialog box.

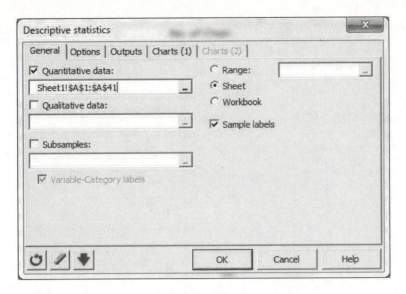

- **Quantitative data**: Enter the data range, **A1:A41**.
- **Sheet**: The output will be displayed in a new Excel worksheet.
- **Sample labels**: Because the top cell in the data range is a label, you need to select **Sample labels**.

5. Click the **Options** tab. Select **Charts**.
6. Click the **Charts(1)** tab. XLSTAT provides several chart options. For this problem, select **Boxplots**.

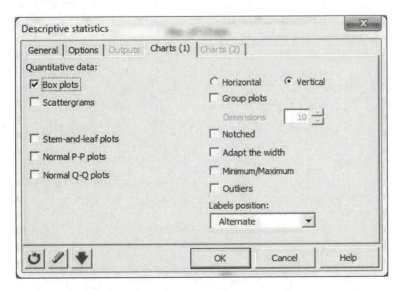

7.   Click **OK**. Click **Continue** in the XLSTAT-Selections dialog box.

The boxplot constructed by XLSTAT is shown below.

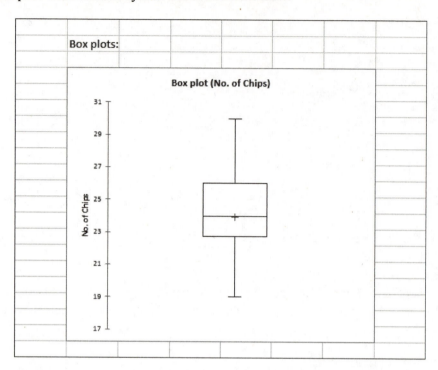

4

Probability

4-6   Counting

| Example 2, page 177 | **Using the FACT Function to Find the Factorial of Six** |

Example 2 asks you to find the probability that a student who makes random guesses will select the correct chronological order of six presidents. You will use Excel's FACT function to calculate 6!.

1. At the top of the screen, click **FORMULAS**. Then click **Insert Function**.
2. Select the **Math & Trig** category. Select the **FACT** function. Click **OK**.

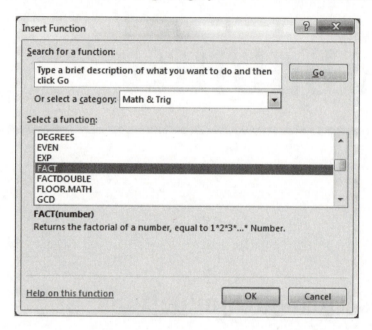

3. Complete the dialog box as shown below. Click **OK**.

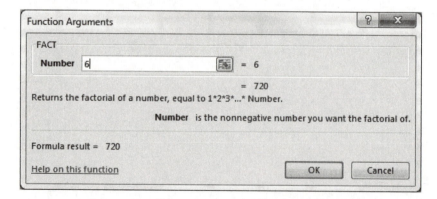

4. Excel's FACT function returns 720. To find the solution to the problem presented in Example 2, you divide 1 by 720: $\frac{1}{720} = 0.00139$.

<table>
<tr><td>Example 3, page 177</td><td>**Using the PERMUT Function to Find the Probability of Randomly Selecting Two Horses in the Correct Order**</td></tr>
</table>

Example 3 asks you to find the probability that a bettor who makes random guesses will select the correct horses to finish first and second out of a total of 20 horses. You will use Excel's PERMUT function to calculate the number of different possible arrangement of 2 horses selected from 20.

1. At the top of the screen, click **FORMULAS**. Then click **Insert Function**.
2. Select the **Statistical** category. Select the **PERMUT** function. Click **OK**.

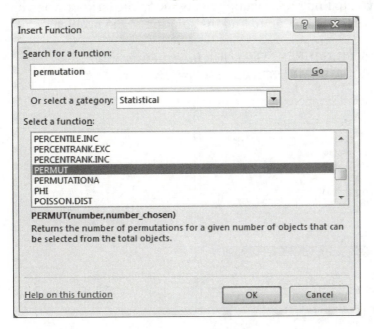

3. Complete the dialog box as shown below. Click **OK**.

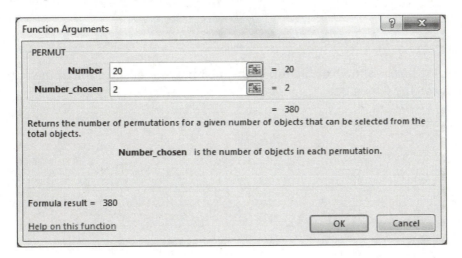

Excel's PERMUT function returns 380. The probability of randomly selecting the winning arrangement is $\frac{1}{380} = 0.00263$.

> **Example 5, page 178**
>
> # Using the COMBIN Function to Find the Probability of Selecting the Correct Combination from 49 Different Numbers

Winning the jackpot requires that you select six winning numbers, drawn in any order, from the numbers 1 to 49. Example 5 asks you to find the probability of winning the jackpot when one ticket is purchased. You will use Excel's COMBIN function to find the solution to the problem.

1. At the top of the screen, click **FORMULAS**. Then click **Insert Function**.
2. Select the **All** category. Select the **COMBIN** function. Click **OK**.

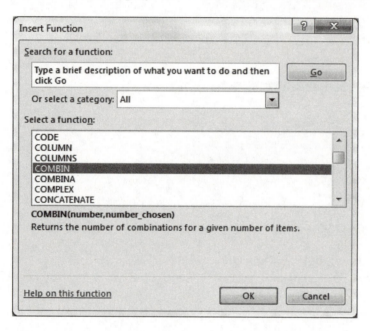

3. Complete the dialog box as shown at the top of the next page. A description of the entries immediately follows the dialog box.

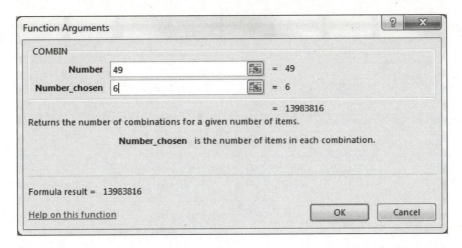

- **Number**: Enter the total number of items. For this problem, the total is **49**.
- **Number_chosen**: Enter the number of items in each combination. For this problem, there are **6** numbers in each combination.

4. Click **OK**. The function returns 13,983,816 combinations. The probability of winning the jackpot is $\frac{1}{13,983,816}$ = 7.15112E-08. Because the probability is so small, it is displayed in scientific notation.

4-7 Probabilities Through Simulations (Technology Project)

Technology Project, page 190

Using the RANDBETWEEN Function to Find the Probability That Two Randomly Selected People Have the Same Birthday

The classic birthday problem presented in the Technology Project asks you to find the probability that among 25 randomly selected people at least two have the same birthday. You will be generating the samples using Excel's **RANDBETWEEN** function.

1. Open a new Excel worksheet and click in cell **A1**.
2. At the top of the screen, click **FORMULAS**. Then click **Insert Function**.
3. Select the **Math & Trig** category. Select the **RANDBETWEEN** function. Click **OK**.
4. In the dialog box, enter **1** for the bottom and **365** for the top as shown at the top of the next page. Click **OK**.

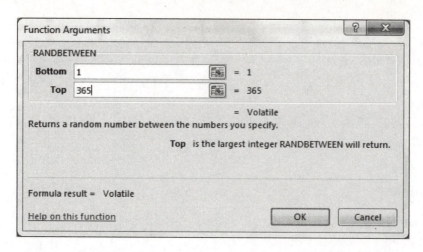

2. Copy the contents of cell A1 to **A2:A25**. The first few rows of my output are shown below. Your output will not be the same as mine because the numbers were generated randomly.

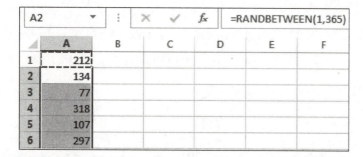

You will notice that the number originally displayed in A1 changed after you executed the copy. This occurred because cell A1 contains the RANDBETWEEN function rather than a numerical value. If you click in any cell in the range A1:A25, you will see that the formula bar displays =RANDBETWEEN(1,365).

3. To prevent the cell contents from changing again, you will want to replace the cell contents with numerical values. To do this, first click and drag over **A1:25** to select the 25 randomly generated numbers. **Right-click** and select **Copy**. **Right-click** again and select **Paste Special**. Select **Values**. Click **OK**.

4. The RANDBETWEEN function in the cells A1:A25 has now been replaced with numerical values. Sort the values in A1:A25 to see if there are any repetitions. My sample, shown below, did not have any repetitions. In order to obtain a good estimate of the probability of two or more persons having the same birthday, you might want to general several more samples of size *n* = 25.

	A
1	43
2	56
3	61
4	77
5	79
6	82
7	92
8	95
9	102
10	104
11	107
12	125
13	134
14	157
15	164
16	212
17	218
18	220
19	226
20	256
21	259
22	297
23	300
24	318
25	340

5

Discrete Probability Distributions

5-3   Binomial Probability Distributions

Example 2, page 213	Using the BINOM.DIST Function to Find the Probability That a Randomly Selected Adult Knows What Twitter Is

Example 2 tells you that there is a 0.85 probability that a randomly selected adult knows what Twitter is. You are asked to use the binomial probability formula to find the probability of getting exactly three adults who know what Twitter is when five adults are randomly selected. You will use Excel's BINOM.DIST function to find the answer.

1. At the top of the screen, click **FORMULAS**. Then click **Insert Function**.
2. Select the **Statistical** category. Select the **BINOM.DIST** function. Click **OK**.

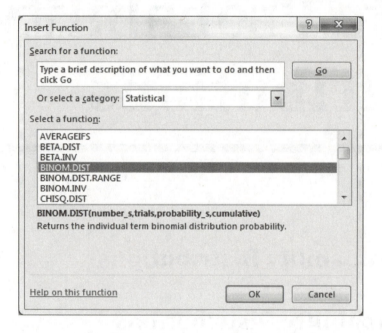

3. Complete the dialog box as shown at the top of the next page. A description of the entries is given immediately after the dialog box.

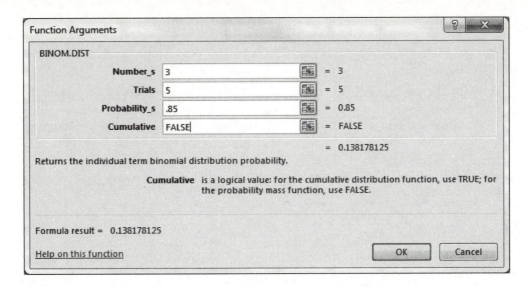

- **Number_s**: The number of successes in trials for this example is **3**.
- **Trials**: The total number of independent trials is the number of random selections. For this example, the number of random selections is **5**.
- **Probability_s**: The probability of success on each trial is **.85**.
- **Cumulative**: Enter TRUE if you would like the cumulative distribution. Enter FALSE if you would like the mass probability, which, for this example, is the probability that there are exactly three successes. For this problem, enter **FALSE**.

4. Click **OK**. The function returns a probability of 0.1382.
5. If you would like to display the probability of all outcomes (i.e., zero to five adults who know what Twitter is), first enter 0 to 5 in column A as shown below. Then click in cell **B1** where the probability of obtaining zero adults will be placed.

	A	B
1	0	
2	1	
3	2	
4	3	
5	4	
6	5	

6. At the top of the screen, click **FORMULAS**. Then click **Insert Function**.
7. Select the **Statistical** category. Select the **BINOM.DIST** function. Click **OK**.
8. Complete the dialog box as shown at the top of the next page. Instead of entering 0 for the number of successes in trials, you enter **A1**, the cell address of zero.

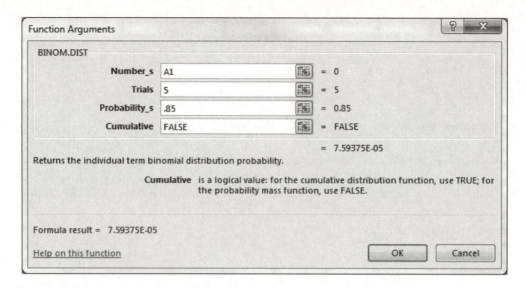

9.   Click **OK**. The function returns a probability of 7.59375E-05.
10. Copy the contents of cell B1 to **B2:B6**.

The probability distribution obtained by using Excel's BINOM.DIST function is displayed below.

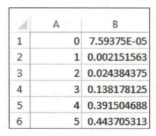

	A	B
1	0	7.59375E-05
2	1	0.002151563
3	2	0.024384375
4	3	0.138178125
5	4	0.391504688
6	5	0.443705313

Example 4, page 215   **Using the BINOM.DIST Function to Find the Probability That a Randomly Selected Adult Believes in the Devil**

Example 4 states that, according to the results of a recent poll, 60% of adults believe in the devil. If five adults are randomly selected, part b of Example 4 asks you to find the probability that at least two believe in the devil. You will be using Excel's BINOM.DIST function to obtain the mass and cumulative probabilities associated with $P(0)$ to $P(5)$.

1.   Begin by entering 0 to 5 in column A as shown at the top of the next page. Then click in cell **B1** where the probability of randomly selecting exactly zero adults who believe in the devil will be placed.

	A	B
1	0	
2	1	
3	2	
4	3	
5	4	
6	5	

2. At the top of the screen, click **FORMULAS**. Then click **Insert Function**.
3. Select the **Statistical** category. Select the **BINOM.DIST** function. Click **OK**.
4. Complete the dialog box as shown at the below. A description of the entries immediately follows the dialog box.

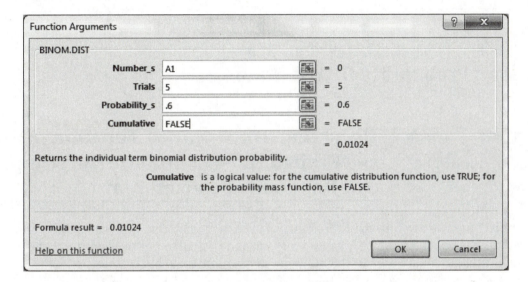

- **Number_s**: The number of successes in trials for this probability distribution is 0 to 5. You are first finding $P(0)$. Rather than entering 0, you enter **A1**, the cell address of 0. You enter the cell address rather than the actual number of successes because this will enable you to copy the function to obtain the probabilities associated with the other outcomes.
- **Trials**: The total number of independent trials is the number of random selections. For this example, the number of random selections is **5**.
- **Probability_s**: The probability of success on each trial is **.60**.
- **Cumulative**: Enter **FALSE** to obtain the probability associated with obtaining exactly zero adults who believe in the devil.

5. Click **OK**. The function returns a probability of 0.0102.
6. Copy the contents of cell B1 to **B2:B6**. These are the mass probabilities.
7. You will next obtain the cumulative probabilities. Click in cell **C1** where the probability of randomly selecting at least 0 adults who believe in the devil will be placed.
8. Repeat steps 2 and 3 shown above. Complete the dialog box as shown in step 4 except enter **TRUE** instead of FALSE in the Cumulative window.
9. Click **OK**. Copy the contents of cell C1 to **C2:C6**.

The output from Excel's BINOM.DIST function is displayed below. To obtain P(at least 2 believe in the devil), you can add the probabilities of obtaining exactly 2, 3, 4, or 5 successes: $P(2) + P(3) + P(4) + P(5) = 0.9130$. You can obtain the same result by subtracting the cumulative probability, P(At least 1 believe in the devil) from 1: $1 - 0.08704 = 0.9130$.

	A	B	C
1	0	0.01024	0.01024
2	1	0.0768	0.08704
3	2	0.2304	0.31744
4	3	0.3456	0.66304
5	4	0.2592	0.92224
6	5	0.07776	1

5-5   Poisson Probability Distributions

Example 1, page 229	Using the POISSON.DIST Function to Find the Probability of x Atlantic Hurricanes in a Randomly Selected Year

Part b of Example 1 asks you to find the probability of zero, two, and nine hurricanes in a randomly selected year. Based on data from the University of Maryland Department of Geography and Environmental Systems, you are able to calculate the mean number of hurricanes per year, $\mu = \frac{530}{100} = 5.3$. You will use Excel's POISSON.DIST function to find the solution to the problem. We will start with $P(0)$.

1. At the top of the screen, click **FORMULAS**. Then click **Insert Function**.
2. Select the **Statistical** category. Select the **POISSON.DIST** function. Click **OK**.
3. Complete the dialog box as shown at the top of the next page. A description of the entries immediately follows the dialog box.

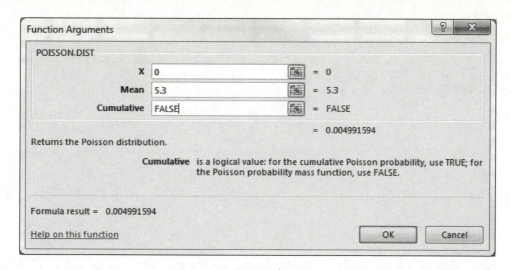

- **X**: X is the number of events. For this example, the number of events is **0** hurricanes.
- **Mean**: Mean is the mean of the Poisson distribution. For this example, $\mu = 5.3$.
- **Cumulative**: If you want the cumulative Poisson probability, you enter TRUE. If you want the mass Poisson probability, you enter FALSE. For this problem, enter **FALSE**.

4. Click **OK**. The function returns 0.00499.
5. To obtain $P(2)$ and $P(9)$, repeat steps 1-3 entering **2** and then **9** for X in the dialog box. The POISSON.DIST function returns 0.0701 and 0.0454, respectively.

6

Normal Probability Distributions

6-2   The Standard Normal Distribution

| **Example 3, page 248** | **Using the NORM.S.DIST Function to Find the Probability of a Bone Density Test Result** |

Example 2 asks you to find the probability that the result of a bone density test is a reading less than 1.27. You will use Excel's NORM.S.DIST function to find the answer.

1. At the top of the screen, click **FORMULAS**. Then click **Insert Function**.
2. Select the **Statistical** category. Select the **NORM.S.DIST** function. Click **OK**.

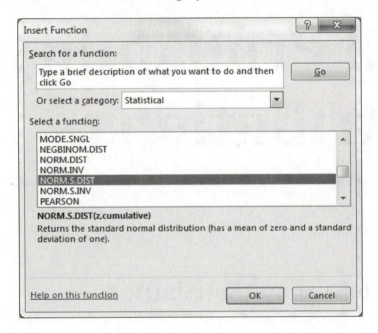

3. The result is measured as a z score. For this example, the z score is equal to **1.27**. Enter **true** for the cumulative distribution.

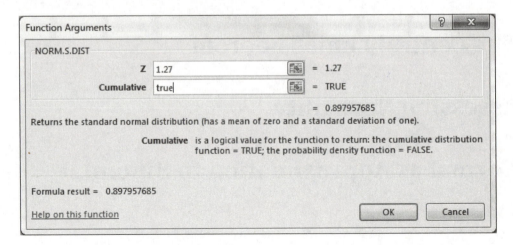

4. Click **OK**. The NORM.S.DIST function returns a probability of 0.8980.

Example 4, page 250

Using the NORM.S.DIST Function to Find the Probability of a Bone Density Test Result

Example 2 asks you to find the probability that the result of a bone density test is a reading above -1.00. You will use Excel's NORM.S.DIST function to find the answer.

1. At the top of the screen, click **FORMULAS**. Then click **Insert Function**.
2. Select the **Statistical** category. Select the **NORM.S.DIST** function. Click **OK**.
3. The result is measured as a z score. For this example, the z score is equal to **–1.00**. Enter **true** for the cumulative distribution.

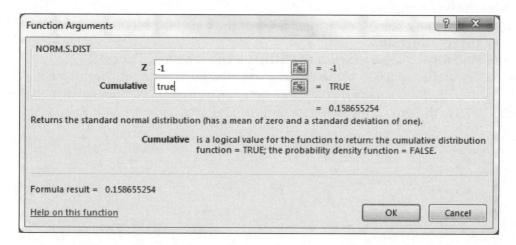

4. Click **OK**. Excel's NORM.S.DIST function returns a probability of 0.1587. This is the area below $z = -1.00$. To find the area above -1.00, you need to subtract 0.1587 from 1. The answer is: $1 - 0.1587 = 0.8413$.

Example 7, page 254

Using the NORM.S.INV Function to Find the Bone Density Test z Score for a Given Area Under the Normal Curve

Example 7 asks you to find the bone density test score that separates the bottom 2.5% and to find the bone density test score that separates the top 2.5%. Recall that bone density test scores are expressed as z scores. You will use Excel's NORM.S.INV function to find the answers.

1. We will start with the bottom 2.5%. At the top of the screen, click **FORMULAS**. Then click **Insert Function**.
2. Select the **Statistical** category. Select the **NORM.S.INV** function. Click **OK**.

3. The probability, expressed as a proportion, is 0.025. Enter **.025** in the Probability window of the dialog box.

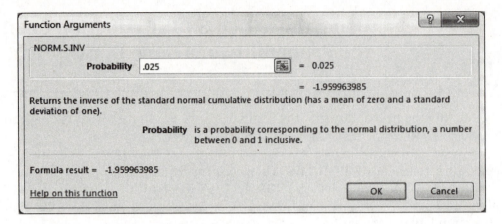

4. Click **OK**. The function returns $z = -1.96$.
5. Next, we will find the z score for the top 2.5%. At the top of the screen, click **FORMULAS**. Then click **Insert Function**.
6. Select the **Statistical** category. Select the **NORM.S.INV** function. Click **OK**.
7. The boundary of the top 0.025 of the distribution is equal to the boundary of the lower $1 - 0.025 = 0.975$. Because the NORM.S.INV function requires a cumulative distribution, you enter **.975** in the Probability window of the dialog box.

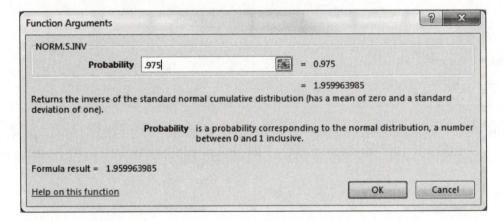

8. Click **OK**. The function returns $z = 1.96$. The boundary of the lower 2.5% is –1.96, and the boundary of the upper 2.5% is 1.96.

6-3 Applications of Normal Distributions

Example 1, page 259	**Using the NORM.DIST Function to Find the Proportion of Women Eligible for Tall Clubs International**

Example 1 says that Tall Clubs International has a requirement that women must be at least 70 in. tall. Given that women have normally distributed heights with a mean of 63.8 in. and standard deviation of 2.6 in., find the percentage of women who satisfy that height requirement. You will use Excel's NORM.DIST function to find the answer.

1. At the top of the screen, click **FORMULAS**. Then click **Insert Function**.
2. Select the **Statistical** category. Select the **NORM.DIST** function. Click **OK**.
3. Complete the dialog box as shown below. Click **OK**.

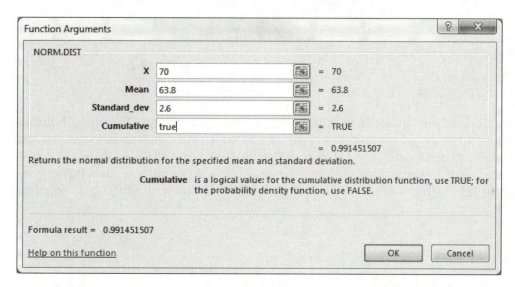

4. Excel's NORM.DIST function returns a cumulative area of 0.9915. This is the area to the left of the point corresponding to 70 in. You want the area to the right of this point. This area is equal to $1 - 0.9915 = 0.0085$. The proportion of women taller than 70 in. is 0.0085.

Example 4, page 264	**Using the NORM.INV Function to Find the Wechsler IQ Scores That Separate Three Groups**

Example 4 says that students are to be placed into three ability groups based on their IQ scores. The first group has scores in the bottom 30%, the second group has scores in the middle 40%, and the third group has scores in the top 30%. You are to find the Wechsler IQ scores that separate the three groups. Wechsler IQ scores are normally distributed and have a mean of 100 and a standard deviation of 15. The answers are the upper and lower boundaries of the middle 40%. You will use Excel's NORM.INV function to find the answers.

1. We will start with the lower boundary of the middle 40%. At the top of the screen, click **FORMULAS**. Then click **Insert Function**.
2. Select the **Statistical** category. Select the **NORM.INV** function. Click **OK**.
3. Complete the dialog box as shown below. Because the probability is cumulative, you enter **.3** to obtain the lower boundary of the middle 40%. Click **OK**.

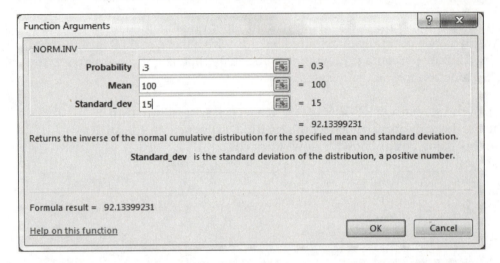

4. The function returns an IQ score equal to 92.13. Next, we will find the upper boundary of the middle 40%. At the top of the screen, click **FORMULAS**. Then click **Insert Function**.
5. Select the **Statistical** category. Select the **NORM.INV** function. Click **OK**.
6. Complete the dialog box as shown below. Because the probability is cumulative, you enter **.7** to obtain the upper boundary. Click **OK**. The function returns an IQ score equal to 107.87.

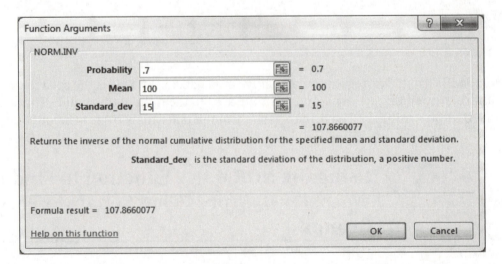

The Wechsler IQ scores that separate the three groups are 92.13 and 107.87.

6-5 The Central Limit Theorem

Example 2, page 286	Using the NORM.DIST Function to Find the Probability That a Randomly Selected Male Weighs More Than 156.25

Example 2 describes an elevator that is loaded to a capacity of 2500 lb. with 16 males. The mean weight of a passenger in this situation is 156.25 lb. You are told that population mean of adult males' weights is 182.9 and the standard deviation is 40.8. You are asked to find the probability that one randomly selected adult male has a weight greater than 156.25. You will use Excel's NORM.DIST function to find the answer.

1. At the top of the screen, click **FORMULAS**. Then click **Insert Function**.
2. Select the **Statistical** category. Select the **NORM.DIST** function. Click **OK**.
3. Complete the dialog box as shown below. Click **OK**.

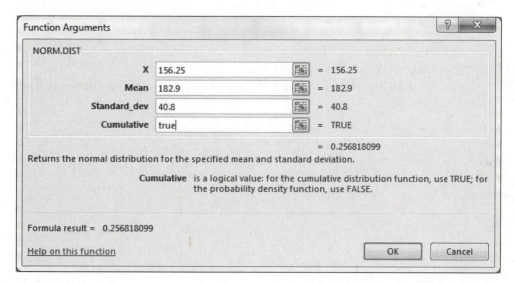

4. Excel's NORM.DIST function returns a probability of 0.2568. Because this is a cumulative distribution function, 0.2568 is the probability of randomly selecting a male with a weight less than 156.25. To find the probability of selecting a male with a weight greater than 156.25, you need to subtract: $1 - 0.2568 = 0.7432$. The probability of randomly selecting a male with a greater than 156.25 is 0.7432.

6-6 Assessing Normality

Example 3, page 300	Using XLSTAT to Assess Normality of Earthquake Magnitudes

Example 3 asks you to assess the normality of the distribution of earthquake magnitudes. You will use XLSTAT's normality tests.

1. Open the **QUAKE** file on your data disk. The data set contains 50 earthquake magnitudes. The first few rows of the data file are shown below.

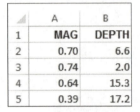

	A	B
1	MAG	DEPTH
2	0.70	6.6
3	0.74	2.0
4	0.64	15.3
5	0.39	17.2

2. At the top of the screen, click **the XLSTAT** add-in.
3. Select **Describing data**. Select **Normality tests**.
4. Complete the General dialog box as shown below. A description of the entries is given immediately after the dialog box.

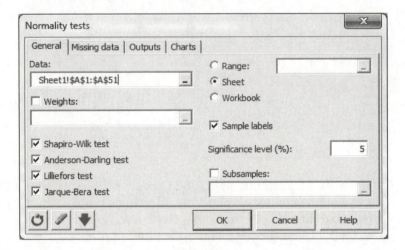

- **Data**: Enter the range of the magnitude data, **A1:A51**.
- **Tests**: XLSTAT offers four different normality tests: **Shapiro-Wilk, Anderson-Darling, Lilliefors, Jarque-Bera**. For this problem, we will select all of them.
- **Sheet**: The output will be placed in a new Excel worksheet.
- **Sample labels**: Because the top cell in the range is a label, you need to select **Sample labels**.
- **Significance level (%)**: Use the default significance level of **5%**.

5. Click the **Charts** tab. Select **Normal Q-Q plots**. Click **OK**. Click **Continue** in the XLSTAT-Selections dialog box.

The results of all four tests state that the null hypothesis cannot be rejected. The XLSTAT output from one of the tests, the Shapiro-Wilk test, is shown below.

Shapiro-Wilk test (MAG):						
W	0.9725					
p-value	0.2925					
alpha	0.05					
Test interpretation:						
H0: The variable from which the sample was extracted follows a Normal distribution.						
Ha: The variable from which the sample was extracted does not follow a Normal distribution.						
As the computed p-value is greater than the significance level alpha=0.05, one cannot reject the null hypothesis H0.						
The risk to reject the null hypothesis H0 while it is true is 29.25%.						

The output from the XLSTAT Q-Q plot is shown below.

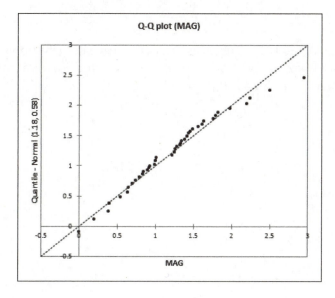

6-7   Normal as Approximation to Binomial

Example 1, page 307	**Using the NORM.DIST Function to Find the Probability of at Least 235 Wins Among 431 Teams Winning the Coin Toss**

Example 1 presents a problem regarding the fairness of the NFL coin toss. Your text explains how to calculate μ (215.5) and σ (10.380270) and also explains why 234.5 is used to calculate the probability. These values will be entered into Excel's NORM.DIST function.

1. At the top of the screen, click **FORMULAS**. Then click **Insert Function**.
2. Select the **Statistical** category. Select the **NORM.DIST** function. Click **OK**.
3. Complete the dialog box as shown below. Click **OK**.

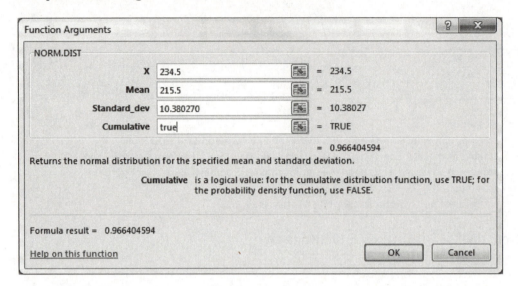

4. Excel's NORM.DIST function returns 0.9664. Because this is a cumulative distribution function, 0.9664 is the area to the left of 235. We want the area to the right. To obtain the area to the right, we need to subtract: $1 - 0.9664 = 0.0336$.

7

Estimates and Sample Sizes

7-2   Estimating a Population Proportion

Example 2, page 328	Using the NORM.S.INV Function to Find a Critical *z* Value

Example 2 asks you to find the critical value $z_{\alpha/2}$ corresponding to a 95% confidence level. You will be using Excel's NORM.S.INV function. The NORM.S.INV function returns the inverse of the standard normal cumulative distribution. As stated in the text, the 95% confidence level corresponds to $\alpha/2 = 0.025$. Therefore, the boundary of the lower tail is 0.025, and the boundary of the upper tail is $1 - 0.025 = 0.975$.

1. At the top of the screen, click **FORMULAS**. Then click **Insert Function**.
2. Select the **Statistical** category. Select the **NORM.S.INV** function. Click **OK**.

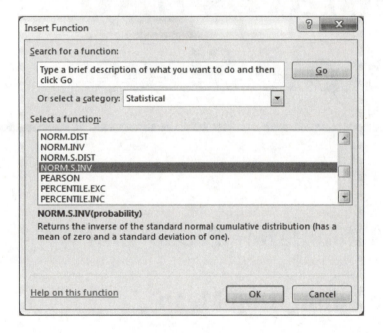

3. For the lower boundary, enter a probability of **0.025**. Click **OK**.

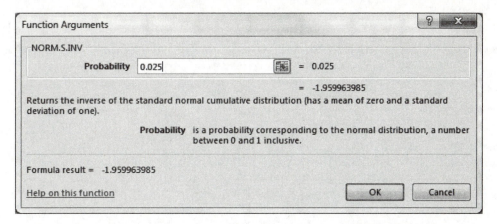

4. The NORM.S.INV function returns –1.96. Select the NORM.S.INV function again. This time enter a probability of **0.975**. Click **OK**. The function returns 1.96.

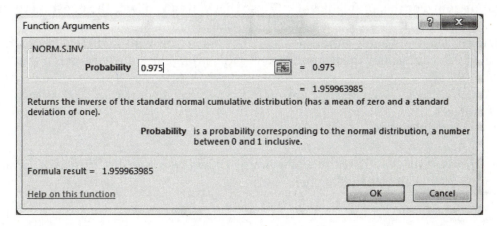

Expressed as an absolute value, a 95% confidence level results in a critical value of $z_{\alpha/2} = 1.96$.

Example 3, page 331   **Using XLSTAT to Construct a Confidence Interval for Poll Results**

Example 3 presents the results of a poll regarding familiarity with Twitter and asks you to find the 95% confidence interval estimate of the population proportion p. The sample results are $n = 1007$ and $\hat{p} = 0.85$. You are also asked, based on the results, whether or not we can safely conclude that more than 75% of adults know what Twitter is. You will use XLSTAT's tests for one proportion to find the answer.

1. At the top of the screen, click **XLSTAT** add-in.
2. Select **Parametric tests**. Select **Tests for one proportion**.
3. Complete the dialog box as shown at the top of the next page. A description of the entries is given immediately after the dialog box.

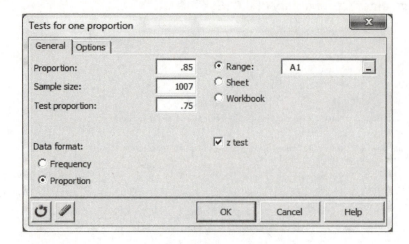

- **Proportion**: Enter the sample result **.85**.
- **Sample size**: Enter the sample size **1007**.
- **Test proportion**: For this problem, enter **.75** so that you can answer the question regarding a conclusion that more than 75% of adults know what Twitter is.
- **Data format**: You could enter either the frequency of persons who said they know what Twitter is or you could enter the proportion. For this problem, you have entered the proportion (0.85).
- **Range**: You are choosing to place the output in the current sheet with cell **A1** as the rightmost top cell.
- **z test**: Select **z test** for the analysis.

4. Click the **Options** tab. Complete the dialog box as shown below. A description of the entries is given immediately after the dialog box.

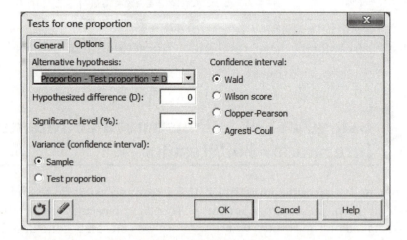

- **Alternative hypothesis**: Select the two-tailed option, **Proportion – Test proportion ≠ D**.
- **Hypothesized difference (D)**: Use the default value **0**.
- **Significance level (%)**: Use the default value **5**.
- **Variance (confidence interval)**: Because you are forming a confidence interval around a sample estimate, you want the variance to be computed using the **Sample** proportion.
- **Confidence interval**: You are given four options for constructing the confidence interval. Select **Wald**.

5. Click **OK**.

The confidence interval portion of the XLSTAT output is shown below. The 95% confidence interval is $0.828 < p < 0.872$. Because the limits of the confidence interval are greater than 0.75, we can safely conclude that more than 75% of adults know what Twitter is.

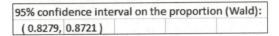

95% confidence interval on the proportion (Wald):		
(0.8279,	0.8721)	

7-3   Estimating a Population Mean

Example 1, page 345	## Using the T.INV.2T Function to Find a Critical *t* Value

Example 1 asks you to find the critical value $t_{\alpha/2}$ for a sample of size $n = 12$ selected from a normally distributed population. You will use Excel's T.INV.2T function. The T.INV.2T function returns the two-tailed inverse of the Student's t-distribution for a specified probability and degrees of freedom.

1. At the top of the screen, click **FORMULAS**. Then click **Insert Function**.
2. Select the **Statistical** category. Select the **T.INV.2T** function. Click **OK**.

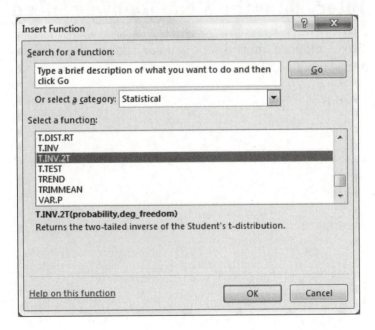

3. Complete the dialog box as shown below. A description of the entries is given immediately after the dialog box.

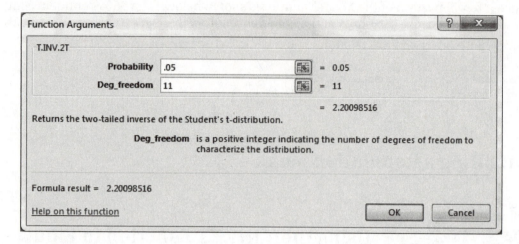

4. **Probability**: The probability is α, and for this problem, $\alpha = .05$.
5. **Deg_freedom**: Degrees of freedom are equal to $n-1 = 12-1 = 11$. Enter **11** for degrees of freedom.

5. Click **OK**. Excel's T.INV.2T function returns critical $t = 2.2010$.

<table>
<tr><td>**Example 2, page 347**</td><td># Using XLSTAT to Construct a Confidence Interval for the Mean Highway Speed</td></tr>
</table>

Example 2 asks you to construct a 95% confidence interval for the mean speed using sample data of size $n = 12$. You will use XLSTAT's one-sample t-test.

1. Enter the data in an Excel worksheet as shown below.

◢	A
1	mi/h
2	62
3	61
4	61
5	57
6	61
7	54
8	59
9	58
10	59
11	69
12	60
13	67

2. At the top of the screen, select the **XLSTAT** add-in.

3. Select **Parametric tests**. Select **One-sample t-test and z-test**.
4. Complete the General dialog box as shown below. A description of the entries is given immediately after the dialog box.

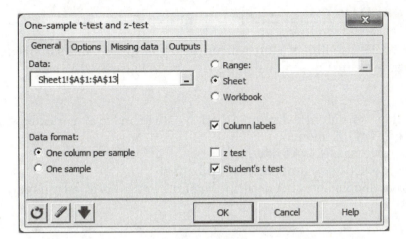

- **Data**: Enter the worksheet range of the data, **A1:A13**.
- **Data format**: Select **One column per sample**.
- **Sheet**: The output will be placed in a new Excel worksheet.
- **Column labels**: Because the top cell of the data range is a label and not a data value, you need to select **Column labels**.
- **Student's t test**: The population variance is unknown and will be estimated from the sample data. The appropriate test is the **Student's t test**.

5. Click the **Options** tab. Select the two-tailed alternative hypothesis. Enter **65** for the theoretical mean. For the significance level, select the default of **5**.

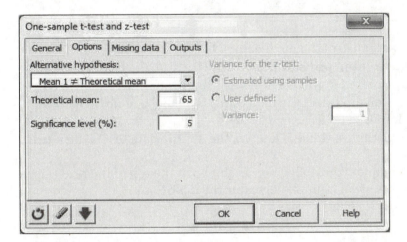

6. Click **OK**. Click **Continue** in the XLSTAT-Selections dialog box.

The confidence interval portion of the output is displayed below. The lower limit of the 95% confidence interval is 58.0775 and the upper limit is 63.2558.

95% confidence interval on the mean:		
(58.0775,	63.2558)	

Example 7, page 353

Using XLSTAT to Construct a Confidence Interval Estimate of μ Assuming σ Is Known

Example 7 asks you to construct a 95% confidence interval for the mean speed using sample data given in Example 2. You are to assume that σ is known to be 4.1. You will use XLSTAT's one-sample z-test.

1. If you have not already done so, enter the Example 2 mi/h data in an Excel worksheet as shown earlier in Section 7-3 of this manual.
2. At the top of the screen, select the **XLSTAT** add-in.
3. Select **Parametric tests**. Select **One-sample t-test and z-test**.
4. Complete the dialog box as shown below. A description of the entries is given immediately after the dialog box.

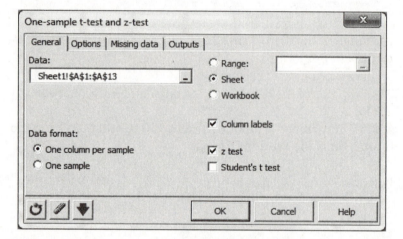

- **Data**: Enter the worksheet range of the mi/h data, **A1:A13**.
- **Data format**: Select **One column per sample**.
- **Sheet**: The output will be placed in a new Excel worksheet.
- **Column labels**: Because the top cell of the data range is a label and not a data value, you need to select **Column labels**.
- **z test**: Because the population variance is known, the appropriate test is the **z-test**.

5. Click the **Options** tab. Complete the dialog box as shown at the top of the next page. A description of the entries is given immediately after the dialog box.

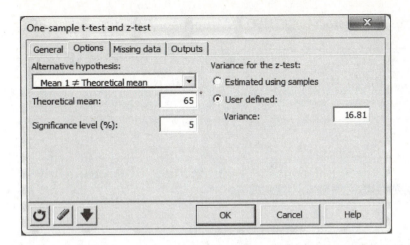

- **Alternative hypothesis**: Select the two-tailed alternative hypothesis.
- **Theoretical mean**: Set the theoretical mean equal to **65**. We are using 65 here because Example 2 asked what the confidence interval suggested regarding the speed limit of 65.
- **Significance level**: Use the default significance level of **5**.
- **Variance for the z-test**: Select **User defined**. You are told that $\sigma = 4.1$. Therefore, the variance is equal to $\sigma^2 = 4.1^2 = 16.81$.

6. Click **OK**. Click **Continue** in the XLSTAT-Selections dialog box.

The confidence interval portion of the XLSTAT output is shown below. The lower limit of the 95% confidence interval is 58.347 and the upper limit is 62.986.

One-sample z-test / Two-tailed test:			
95% confidence interval on the mean:			
(58.3469,	62.9864)		

7-4 Estimating a Population Standard Deviation or Variance

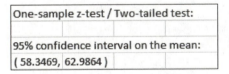

Example 1, page 363 — **Using the CHISQ.INV and CHISQ.INV.RT Functions to Find a Critical Value of X²**

Example 1 asks you to find the critical value of X^2 separating an area of 0.025 in the left tail and to find the critical value of X^2 separating an area of 0.025 in the right tail. The number of degrees of freedom is $df = n - 1 = 22 - 1 = 21$. You will be using Excel's CHISQ.INV and CHISQ.INV.RT functions.

1. We will start with the critical value for the left tail. At the top of the screen, click **FORMULAS**. Then click **Insert Function**.

2. Select the **Statistical** category. Select the **CHISQ.INV** function. Click **OK**.

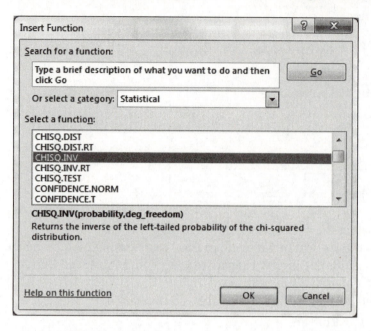

3. The probability is the area in the left tail, **0.025**. Degrees of freedom are equal to **21**. Enter these values in the dialog box as shown below. Click **OK**. Excel's CHISQ.INV function returns critical $X^2 = 10.2829$.

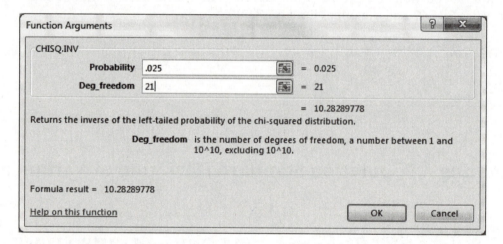

4. Next, we'll find the critical value for the right tail. At the top of the screen, click **FORMULAS**. Then click **Insert Function**.
5. Select the **Statistical** category. Select the **CHISQ.INV.RT** function. Click **OK**.

6. Complete the CHISQ.INV.RT dialog box as shown below. A description of the entries immediately follows the dialog box.

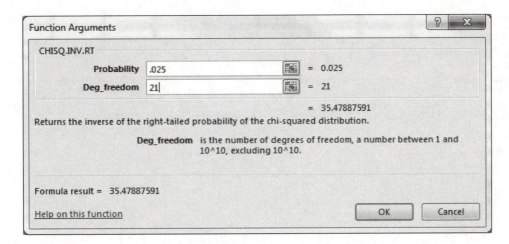

- **Probability**: The probability is the area in the left tail, **.025**.
- **Deg_freedom**: Degrees of freedom are equal to **21**.

7. Click **OK**. Excel's CHISQ.INV.RT function returns critical $X^2 = 35.4789$.

The critical chi-square that separates an area of 0.025 in the left tail is 10.2829. The critical chi-square that separates an area of 0.025 in the right tail is 35.4789.

8

Hypothesis Testing

8-2 Basics of Hypothesis Testing

Example 1, page 383	Using XLSTAT to Test the Claim That the XSORT Gender-Selection Method Is Effective

Example 1 states that couples treated with the XSORT method of gender selection are claimed to be more likely to have a girl than a boy. If 58 of 100 babies are girls, test the claim that $p > 0.5$ where p is the proportion of girls born with the XSORT method. To test this claim, you will use XLSTAT's tests for one proportion.

1. At the top of the screen, select the **XLSTAT** add-in.
2. Select **Parametric tests**. Select **Tests for one proportion**.
3. Complete the General dialog box as shown below. A description of the entries immediately follows the dialog box.

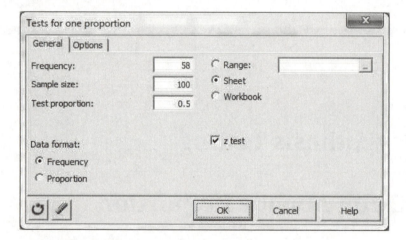

- **Frequency**: The frequency for this problem is **58**.
- **Sample size**: The sample size is **100**.
- **Test proportion**: You are testing the claim that the proportion of girls is greater than **0.5**.
- **Data format**: The format you are using for entering values in the dialog box is **Frequency**.
- **Sheet**: The output will be placed in a new Excel worksheet.
- **z test**: You would like the output to include the results of the **z-test** for one proportion.

4. Click the **Options** tab. Complete the Options dialog box as shown at the top of the next page. A description of the entries immediately follows the dialog box.

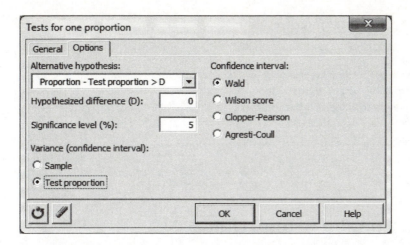

- **Alternative hypothesis**: Select the alternative hypothesis that states *Proportion – Test proportion > D*.
- **Significance level**: Select a **5%** significance level.
- **Variance (confidence interval)**: Because you are carrying out a hypothesis test, the appropriate variance for the confidence interval is the **Test proportion**. The test proportion for this problem is 0.5.
- **Confidence interval**: XLSTAT offers four procedures for constructing the confidence interval. Select **Wald**.

5. Click **OK**.

The XLSTAT output is shown below. The first part of the output displays the values that you entered for frequency, sample size, and test proportion. Also displayed are your selections for the hypothesized difference, the variance for the confidence interval, and the significance level.

Frequency: 58			
Sample size: 100			
Test proportion: 0.5			
Hypothesized difference (D): 0			
Variance (confidence interval): Test proportion			
Significance level (%): 5			

The next part displays the confidence interval and the results of a *z*-test for one proportion. The test results include the observed difference (0.58 – 0.50 = 0.08), the observed value of *z*, the critical value of *z*, one-tailed *p*-value associated with observed *z*, and alpha.

95% confidence interval on the proportion (Wald):			
(0.4820,	0.6780)		
z-test for one proportion / Upper-tailed test:			
Difference	0.0800		
z (Observed value)	1.6000		
z (Critical value)	1.6449		
p-value (one-tailed)	0.0548		
alpha	0.05		

The final part of the XLSTAT output is a test interpretation.

Test interpretation:				
H0: The difference between the proportions is equal to 0.				
Ha: The difference between the proportions is greater than 0.				
As the computed p-value is greater than the significance level alpha=0.05, one cannot reject the null hypothesis H0.				
The risk to reject the null hypothesis H0 while it is true is 5.48%.				

8-3   Testing a Claim About a Proportion

Example 1, page 400

Using XLSTAT to Test the Claim That 93% of Computers Have Antivirus Programs

Section 8-3 in the Triola text states that you can test a claim using one of three methods: *P*-value method, critical value method, or confidence interval method. The output from XLSTAT's tests for one proportion provides information for testing a claim by all three methods. You will be using the data in Example 1 on page 400 of the Triola text. The problem states that, in a random sample of 400 scanned computers, it was found that 380 of them (95%) had antivirus programs. The instructions ask you to use the sample data from the scanned computers to test the claim that 93% of computers have antivirus programs.

1.   At the top of the screen, select the **XLSTAT** add-in.
2.   Select **Parametric tests**. Select **Tests for one proportion**.
3.   Complete the General dialog box as shown below. A description of the entries immediately follows the dialog box.

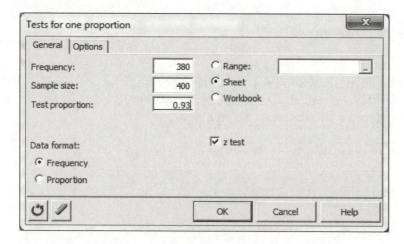

- **Frequency**: The frequency for this problem is **380**.
- **Sample size**: The sample size is **400**.
- **Test proportion**: You are testing the claim that the proportion is equal to **0.93**.

- **Data format**: The format you are using for entering values in the dialog box is **Frequency**.
- **Sheet**: The output will be placed in a new Excel worksheet.
- **z test**: You would like the output to include the results of the **z-test** for one proportion.

4. Click the **Options** tab. A description of the entries immediately follows the dialog box.

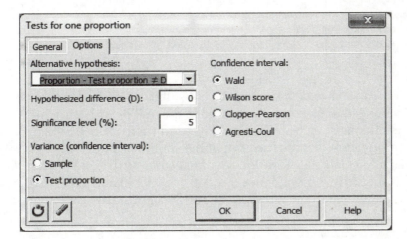

- **Alternative hypothesis**: Select the two-tailed alternative hypothesis that states **Proportion – Test proportion ≠D**.
- **Significance level**: Select a **5%** significance level.
- **Variance (confidence interval)**: Because you are carrying out a hypothesis test, the appropriate variance for the confidence interval is the **Test proportion**. The test proportion for this problem is 0.93.
- **Confidence interval**: XLSTAT offers four procedures for constructing the confidence interval. Select **Wald**.

5. Click **OK**.

The XLSTAT output for testing a claim using each of the three methods (*P*-value method, critical value method, and confidence interval method) is shown below and at the top of the next page.

P-value method. The information needed to determine statistical significance using the *p*-value method is displayed in the XLSTAT output section labeled "*z*-test for one proportion/Two-tailed test." The two-tailed *p*-value is equal to 0.1169. Because the *p*-value is greater than alpha (0.05), the result is not statistically significant.

z-test for one proportion / Two-tailed test:		
Difference	0.0200	
z (Observed value)	1.5677	
z (Critical value)	1.9600	
p-value (Two-tailed)	0.1169	
alpha	0.05	

Critical value method. The information needed to determine statistical significance using the critical value method is also displayed in the XLSTAT output section labeled "*z*-test for one proportion/Two-tailed test." The critical *z* value is equal to 1.9600 and the observed *z* value is equal to 1.5677. Because the observed *z* value is less than the critical z value, the result is not statistically significant.

z-test for one proportion / Two-tailed test:		
Difference	0.0200	
z (Observed value)	1.5677	
z (Critical value)	1.9600	
p-value (Two-tailed)	0.1169	
alpha	0.05	

Confidence interval method. When you are using the confidence interval method to test a hypothesis about a proportion, you need to use the value stated in the null hypothesis when computing the variance for the confidence interval. The hypothesized value is 0.93. When making your selections, you chose "test proportion" for the variance of the confidence interval. This means that you chose 0.93, the hypothesized value. Because you set alpha equal to 0.05, a 95% confidence interval is constructed. The limits of the 95% confidence interval are 0.9250 and 0.9750. Because the hypothesized value of 0.93 is contained in the interval, the result is not statistically significant.

Variance (confidence interval): Test proportion		
Significance level (%): 5		
95% confidence interval on the proportion (Wald):		
(0.9250,	0.9750)	

If the result had been statistically significant, then the variance for the confidence interval would be computed using the sample value of 0.95. You would make this selection in the Options dialog box. The result based on this selection is shown below. When the variance is calculated using the sample value, the limits of the 95% confidence interval are 0.9286 and 0.9714.

Variance (confidence interval): Sample		
Significance level (%): 5		
95% confidence interval on the proportion (Wald):		
(0.9286,	0.9714)	

Test interpretation. The final part of the XLSTAT output presents the test interpretation. This interpretation states that you cannot reject the null hypothesis. In other words, the result is not statistically significant.

Test interpretation:			
H0: The difference between the proportions is equal to 0.			
Ha: The difference between the proportions is different from 0.			
As the computed p-value is greater than the significance level alpha=0.05, one cannot reject the null hypothesis H0.			
The risk to reject the null hypothesis H0 while it is true is 11.69%.			

8-4   Testing a Claim About a Mean

Example 1, page 413	**Using XLSTAT to Test the Claim That Cell Phones Have a Mean Radiation Level Less Than 1.00 W/kg**

To test a claim about a mean, you can use the same three methods that were illustrated for testing a claim about a proportion: *P*-value method, critical value method, or confidence interval method. The output from XLSTAT's one-sample *t*-test provides information for testing a claim by all three methods. You will be using the data in Example 1 on page 413 of the Triola text. The instructions state that you are to use a 0.05 significance level to test the claim that cell phones have a mean radiation level that is less than 1.00 W/kg.

1. Enter the sample data in an Excel worksheet as shown below.

	A
1	W/kg
2	0.38
3	0.55
4	1.54
5	1.55
6	0.5
7	0.6
8	0.92
9	0.96
10	1
11	0.86
12	1.46

2. At the top of the screen, select the **XLSTAT** add-in.
3. Select **Parametric tests**. Select **One-sample t-test and z-test**.
4. Complete the General dialog box as shown below. A description of the entries immediately follows the dialog box.

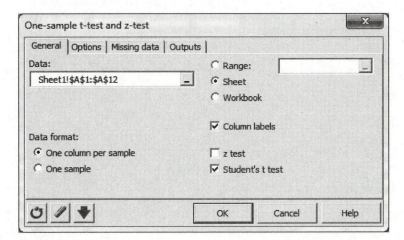

- **Data**: Enter the range of the W/kg data, **A1:A12**.
- **Data format**: Select **One column per sample**.
- **Sheet**: The output will be placed in a new Excel worksheet.
- **Column labels**: Because the top cell in the data range is a label and not a data value, you need to select **Column labels**.
- **Student's t test**: Because the population variance is not known and will be estimated from the sample, the appropriate test is the **Student's *t*-test**.

5. Click the **Options** tab. Complete the dialog box as shown below. A description of the entries immediately follows the dialog box.

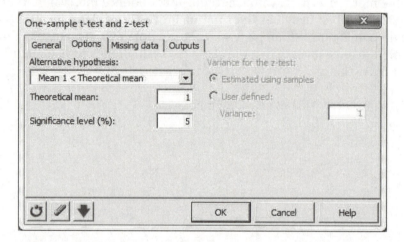

- **Alternative hypothesis**: Select the one-tailed alternative hypothesis that states **Mean 1 < Theoretical mean**.
- **Theoretical mean:** You are testing the claim that cell phones have a mean radiation level that is less than 1 W/kg. Enter **1** for the theoretical mean.
- **Significance level:** Select the default significance level, **5%**.

6. Click the **Outputs tab** and select **Descriptive Statistics**.
7. Click **OK**. Click **Continue** in the XLSTAT-Selections dialog box.

The XLSTAT output for testing a claim using each of the three methods (*P*-value method, critical value method, and confidence interval method) is shown below and explanations are given at the top of the next page.

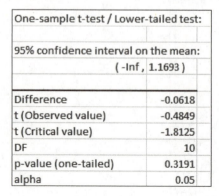

One-sample t-test / Lower-tailed test:	
95% confidence interval on the mean:	
(-Inf , 1.1693)	
Difference	-0.0618
t (Observed value)	-0.4849
t (Critical value)	-1.8125
DF	10
p-value (one-tailed)	0.3191
alpha	0.05

***P*-value method**. The one-tailed *p*-value is equal to 0.3191. Because the *p*-value is greater than alpha (0.05), the result is not statistically significant.

Critical value method. The critical *t* value is equal to –1.8125 and the observed *t* value is equal to –0.4849. Because a lower-tailed test was conducted, the observed *t* must be less than critical *t* in order to be statistically significant. For this problem, observed t is equal to –0.4849 and critical *t* is equal to –1.8125. Because observed t is not less than critical *t*, the result is not statistically significant.

Confidence interval method. The one-tailed confidence interval for a lower-tailed test always has a lower limit of negative infinity. The upper limit of the one-tailed 95% confidence for our example problem is 1.1693. Because the hypothesized value of 1 is contained in the interval, the result is not statistically significant.

Test interpretation. The test interpretation states that you cannot reject the null hypothesis. In other words, the result is not statistically significant.

Test interpretation:			
H0: The difference between the means is equal to 0.			
Ha: The difference between the means is lower than 0.			
As the computed p-value is greater than the significance level alpha=0.05, one cannot reject the null hypothesis H0.			
The risk to reject the null hypothesis H0 while it is true is 31.91%.			

8-5   Testing a Claim About a Standard Deviation or Variance

Example 1, page 425	**Using the CHISQ.INV and CHISQ.DIST Functions to Test a Claim About a Standard Deviation**

Example 1 asks you to test the claim that supermodels have heights with a standard deviation that is less than 2.6 in. for the population of women. Neither Excel's Data Analysis Tools nor XLSTAT includes a statistical test of a standard deviation (or variance) for one sample. However, you can use Excel functions to find the chi-square critical value for the statistical test and the *p*-value of observed chi-square. The CHISQ.INV function will be used to find the critical value, and CHISQ.DIST will be used to find the *p*-value. The CHISQ.DIST function returns the left-tailed probability of the chi-square distribution.

1. We will first find the critical value of chi-square using the CHISQ.INV function. At the top of the screen, click **FORMULAS** and selection **Insert Function**.

2. Select the **Statistical** category and the **CHISQ.INV** function. Click **OK**.

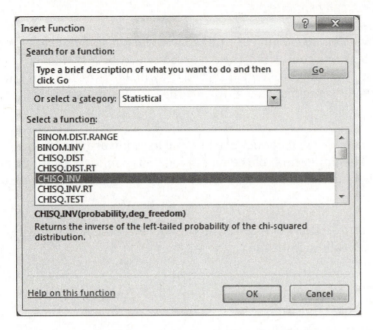

3. Complete the CHISQ.INV dialog box as shown below. A description of the entries immediately follows the dialog box.

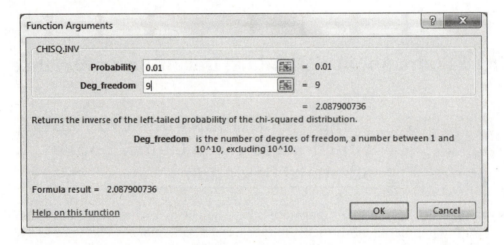

- **Probability:** The probability is the significance level for the test. The instructions in Example 1 tell you to use a significance level of **0.01**.
- **Deg_freedom**: For this problem, n = 10. Degrees of freedom $= n - 1 = 10 - 1 = 9$.

4. Click **OK**. Excel's CHISQ.INV returns 2.0879, the critical value of chi-square for the test.
5. Next, we will find the *p*-value associated with observed chi-square. Observed chi-square is calculated on page 426 of the Triola text. It is equal to 0.852. We need this value for the CHISQ.DIST function. At the top of the screen, click **FORMULAS** and select **Insert Function**.
6. Select the **Statistical** category and the **CHISQ.DIST** function. Click **OK**.
7. Complete the CHISQ.DIST dialog box as shown at the top of the next page. A description of the entries immediately follows the dialog box.

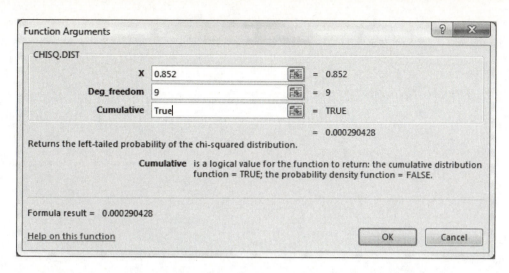

- **X:** X refers to the value of observed chi-square. The observed chi-square is equal to **0.852**.
- **Deg-freedom:** Degrees of freedom are equal to **9.**
- **Cumulative:** Enter **True** to obtain the cumulative distribution function.

8. Click **OK**. Excel's CHISQ.DIST function returns a *p*-value of 0.00029.

Critical value method. The rejection region of the chi-square distribution for this test is in the left tail. The test result is statistically significant if observed chi-square is less than critical chi-square. Critical chi-square is equal to 2.0879, and observed chi-square is equal to 0.852. Therefore, the test result is statistically significant.

P-value method. The significance level of the test (alpha) was set equal to 0.01. The test result is statistically significant if the *p*-value is less than alpha. The p-value of 0.00029 is less than 0.01. Therefore, the test result is statistically significant.

9

Inferences from Two Samples

9-2   Two Proportions

Example 1, page 444	**Using XLSTAT to Test a Claim About Two Proportions**

You will be using the XLSTAT add-in to carry out a z-test that tests the hypothesis that two population proportions are equal to one another. You will need to identify the number of observations in sample 1 (n_1) and the number of successes in sample 1 (x_1). You will also need to identify the number of observations in sample 2 (n_2) and the number of successes in sample 2 (x_2). Because the data are often presented as sample proportions or percentages instead of the actual numbers of successes, you need to know how to find the number of successes from proportions or percentages. By rearranging the terms in $\hat{p}_1 = \dfrac{x_1}{n_1}$, you can solve directly for x_1 by multiplying sample size 1 by the proportion of successes in sample 1. The formula is $x_1 = n_1 \cdot \hat{p}_1$. You will follow the same procedure to solve for the number of successes in sample 2. For example, let's say that sample 1 was comprised of 1,300 individuals and that 45.1% of these individuals responded "yes" to a question on a survey. In this example, $n_1 = 1300$ and $\hat{p}_1 = 0.451$. Solving for x_1, you obtain $x_1 = n_1 \cdot \hat{p}_1 = 1300 \cdot 0.451 = 586.3$ which you round to 586. After identifying n_1, x_1, n_2, and n_2, you are ready to proceed with XLSTAT.

To illustrate how to use XLSTAT to test a claim regarding the equality of two population proportions, we will use the data from Example 1 on page 444 of the Triola text. This example concerns the spending characteristics of people who have larger denominations of money compared to people who have smaller denominations. You are testing the claim that money in large denominations is less likely to be spent relative to an equivalent amount in smaller denominations. You are to test the claim using a 0.05 significance level. In this example, Group 1 has larger denominations of money than Group 2.

1. We begin by identifying x_1, n_1, x_2, and n_2. These values are given in the problem as follows:

   $x_1 = 12$, $n_1 = 46$, $x_2 = 27$, and $n_2 = 43$.

2. At the top of the screen, select the **XLSTAT** add-in.
3. Select **Parametric tests**. Select **Tests for two proportions**.
4. Complete the General dialog box as shown at the top of the next page. A description of the entries immediately follows the dialog box.

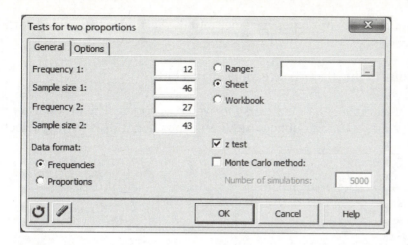

- **Frequency 1**. Enter **12**, the number of successes in sample 1.
- **Sample size 1**: Enter **46**, the size of sample 1.
- **Frequency 2**: Enter **27**, the number of successes in sample 2.
- **Sample size 2**: Enter **43**, the size of sample 2.
- **Data format**: XLSTAT gives you two options for data format. You can either enter frequencies or proportions. For this problem, you are entering **Frequencies**.
- **Sheet**: The output will be placed in a new Excel worksheet.
- **z test**: Because you want the output to include the results of a z-test, you need to select **z-test**.

5. Next, click the **Options** tab and complete the dialog box as shown below. A description of the entries is given immediately after the dialog box.

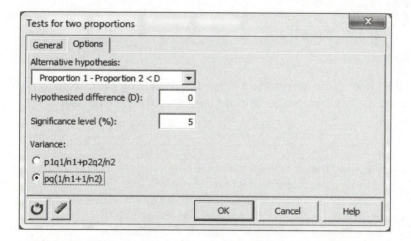

- **Alternative hypothesis**: Three options are provided for the alternative hypothesis. The first is a two-tailed alternative hypothesis. The second is a one-tailed alternative hypothesis with the rejection region in the lower tail. The third is a one-tailed alternative hypothesis with the rejection region in the upper tail. For this problem, you want the second option, **Proportion 1 – Proportion 2 < D**.
- **Significance level (%)**: Use the default significance level, **5%**.

- **Variance**: The first option is the unpooled variance and the second is the pooled variance. You want the pooled variance.

5. Click **OK**. Click **Continue** in the XLSTAT-Selections dialog box.

The output appears in an Excel worksheet labeled *Tests for two proportions*. The first part of the XLSTAT output, shown below, repeats the frequency values and your hypothesis testing selections.

	A	B	C	D
1		XLSTAT 2012.6.04 - Tests for two proportions		
2		Frequency 1: 12		
3		Sample size 1: 46		
4		Frequency 2: 27		
5		Sample size 2: 43		
6		Hypothesized difference (D): 0		
7		Variance: pq(1/n1+1/n2)		
8		Significance level (%): 5		

The second part of the output displays the confidence interval for your chosen alpha, the observed difference, the observed *z*, the critical z, the *p*-value, and alpha.

11		z-test for two proportions / Lower-tailed test:				
12						
13		95% confidence interval on the difference between the proportions:				
14		(-1.0000 , -0.1939)				
15						
16		Difference	-0.3670			
17		z (Observed value)	-3.4874			
18		z (Critical value)	-1.6449			
19		p-value (one-tailed)	0.0002			
20		alpha	0.05			

The final part of the XLSTAT output provides information regarding test interpretation. The one-tailed *z*-test is statistically significant. The researcher should conclude that proportion 1 is significantly less than proportion 2.

22		Test interpretation:
23		H0: The difference between the proportions is equal to 0.
24		Ha: The difference between the proportions is lower than 0.
25		As the computed p-value is lower than the significance level alpha=0.05, one should reject the null hypothesis H0, and accept the alternative hypothesis Ha.
26		The risk to reject the null hypothesis H0 while it is true is lower than 0.02%.

9-3 Two Means: Independent Samples

Three different inference situations will be illustrated for testing the equality of two means when you are dealing with independent samples. The first is the situation where the population standard deviations are not known but are assumed to be unequal. The second situation is one which the population standard deviations are not known but are assumed to be equal. The third is the situation where the population standard deviations are both known. You can use either Excel's Data Analysis Tools or XLSTAT procedures for all three situations. For each situation, I will first explain how to use Excel's Data Analysis Tools. Then I will explain how to use XLSTAT.

Exercise 8, page 464	Using Data Analysis Tools to Test a Claim About Two Means When Standard Deviations Are Unknown and Assumed to Be Unequal

To illustrate how to use Excel's Data Analysis Tools to carry out a *t*-test when population standard deviations are unknown and assumed to be unequal, we will do Exercise 8 on page 464 of the Triola text. This exercise asks you to test the claim that the mean number of words spoken in a day by men is less than the mean number of words spoken in a day by women.

1. Begin by opening the **WORDS** data file on your data disk.
2. The data for both males and females have been entered in six different columns of an Excel worksheet. Your first task is to put all the males' data in a single column and then put all the females' data in a single column. I put the males' data in column A and the females' data in column B. I also included a label in the top cell of each column. The first few rows of the rearranged data set are shown below.

	A	B
1	Male	Female
2	27531	20737
3	15684	24625
4	5638	5198
5	27997	18712
6	25433	12002

3. Click **DATA** at the top of the screen and select **Data Analysis** on the far right of the data ribbon.

If Data Analysis does not appear as a choice in the data ribbon, you will need to load the Microsoft Excel ToolPak add-in. Follow the procedure on page 10.

4. The dialog box presents three t-test options. You want the second one, ***t*-Test: Two-Sample Assuming Unequal Variances**. Click **OK**.

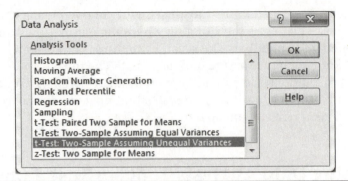

Click the + at the bottom of the screen if the t-test dialog box does not appear.

5. Complete the dialog box as shown below. Detailed descriptions of the entries are given immediately following the dialog box.

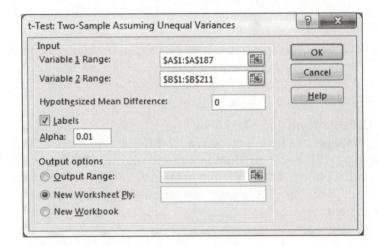

- **Variable 1 Range**: Enter the worksheet location of the males' data, **A1:A187**.
- **Variable 2 Range**: Enter the worksheet location of the females' data, **B1:B211**.
- **Hypothesized Mean Difference**: The default value is zero. It is not necessary to enter anything here if you are using the default value.
- **Labels**: A check mark should appear in the box to the left of Labels to indicate that the top cell in each variable range is a label (Male or Female) and not a data value.
- **Alpha**: The instructions in Exercise 8 ask you to use a 0.01 significance level, so enter **0.01** in the Alpha window.
- **Output options**: You are given three options for placement of the *t*-test output. Let's place the output in a **New Worksheet Ply**.

6. Click **OK**.

The Excel output is displayed below. You will want to make column A wider so that you can see all the output labels. An explanation of the output is given right below the output.

	A	B	C
1	t-Test: Two-Sample Assuming Unequal Variances		
2			
3		*Male*	*Female*
4	Mean	15668.53	16215.03
5	Variance	74520827	53307840
6	Observations	186	210
7	Hypothesized Mean Difference	0	
8	df	364	
9	t Stat	-0.67552	
10	P(T<=t) one-tail	0.249889	
11	t Critical one-tail	2.336636	
12	P(T<=t) two-tail	0.499778	
13	t Critical two-tail	2.589403	

- **Mean**: The sample mean for males is 15,668.53 and the sample mean for females is 16,215.03.
- **Variance**: The variance for the males' data is 74,520,827 and the variance for the females' data is 53,307,840.
- **Observations**: The sample size for males is 186 and the sample size for females is 210.
- **Hypothesized mean difference**: The mean difference stated in the null hypothesis is 0.
- **df**: Degrees of freedom for the test equal 364.
- **t stat:**. The observed value of the t statistic is -0.676.
- **P(T<=t) one-tail**: The one-tailed probability of observed t equals 0.250.
- **t critical one-tail:**. The one-tailed critical value of t for your chosen alpha (0.01) equals 2.337.
- **P(T<=t) two-tail**: The two-tailed probability of observed t equals 0.500.
- **t critical two-tail**: The two-tailed critical value of t for your chosen alpha (0.01) equals 2.589.

Test Interpretation: Because you are carrying out a one-tailed test of the hypothesis, you need to compare the one-tailed p-value to alpha to determine statistical significance. The one-tailed p-value (0.250) is greater than 0.01, the value you selected for alpha. Therefore, you would conclude that the difference is not statistically significant and that the population means of the two groups are equal to one another.

Exercise 8, page 464	# Using XLSTAT to Test a Claim About Two Means When Standard Deviations Are Unknown and Assumed to Be Unequal

To explain how to use XLSTAT to carry out a t-test for the situation of unknown and unequal population standard deviations, we will use the same exercise that we used to explain the use of

Excel's Data Analysis Tools, Exercise 8 on page 464 of the Triola text. You are asked to test the claim that the mean number of words spoken in a day by men is less than that for women.

1. If you have not already done so, complete steps 1 and 2 on page 97 of this manual.
2. At the top of the screen, select the **XLSTAT** add-in.
3. Select **Parametric tests**. Select **Two-sample t test and z test**.
4. Complete the General dialog box as shown below. A description of the entries is given immediately after the dialog box.

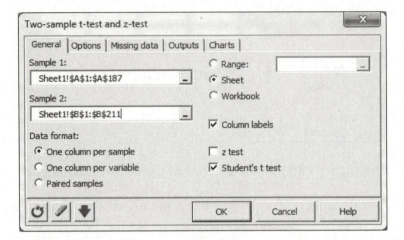

- **Sample 1**: Enter the range of the males' data, **A1:A187**.
- **Sample 2**: Enter the range of the females' data, **B1:B211**.
- **Data format**: Select **One column per sample**.
- **Sheet**: The output will be placed in a new Excel worksheet.
- **Column labels**: Because the top cells of the data ranges contain labels and not data values, you need to select **Column labels**.
- **Student's t test**: Because the population standard deviations are not known and are estimated from the sample data, the appropriate test is the **Student's *t*-test**.

5. Click the **Options** tab at the top of the dialog box. Complete the Options dialog box as shown below. A description of the entries is given immediately after the dialog box.

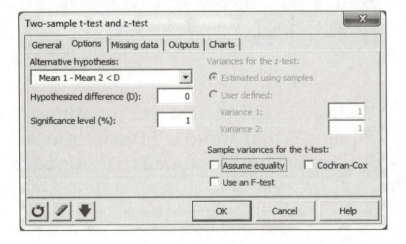

- **Alternative hypothesis**: We want to carry out a one-tailed test where the alternative hypothesis states that the males' mean is less than the females' mean. Because we entered the males' data as sample 1 and the females' data as sample 2, the appropriate alternative hypothesis is **Mean 1 – Mean 2 < D**.
- **Hypothesized difference (D)**: The hypothesized difference is **0**.
- **Significance level**: Enter **1** for the significance level to set the significance level equal to 0.01 (or 1%).
- **Sample variances for the t-test**: You want to carry out the *t*-test for which population variances are assumed to be unequal. To do this, you need to remove the check mark in the box next to **Assume equality**.

6. We don't have any missing data, so we don't need to select an option for dealing with missing data. Click the **Outputs** tab. Make sure that **Descriptive statistics** has been selected.
7. Click **OK**. Click **Continue** in the XLSTAT-Selections dialog box.

The XLSTAT output is shown below. The first part of the output displays descriptive statistics.

Variable	Observations	Obs. with missing data	Obs. without missing data	Minimum	Maximum	Mean	Std. deviation
Male	186	0	186	694.7027	47015.6250	15668.5341	8632.5446
Female	210	0	210	1673.8462	40054.6154	16215.0322	7301.2218

The XLSTAT output labeled *t-test for two independent samples/Lower-tailed test* includes a one-tailed 99% confidence interval and the *t*-test results.

t-test for two independent samples / Lower-tailed test:			
99% confidence interval on the difference between the means:			
(-Inf ,	1343.8567)		
Difference	-546.4981		
t (Observed value)	-0.6755		
t (Critical value)	-2.3366		
DF	364		
p-value (one-tailed)	0.2499		
alpha	0.01		
The number of degrees of freedom is approximated by the Welch-Satterthwaite formula			

The final part of the XLSTAT output presents a test interpretation. For this analysis, the two means are not significantly different.

Test interpretation:			
H0: The difference between the means is equal to 0.			
Ha: The difference between the means is different from 0.			
As the computed p-value is greater than the significance level alpha=0.01, one cannot reject the null hypothesis H0.			
The risk to reject the null hypothesis H0 while it is true is 49.98%.			

Exercise 10, page 464	Using Data Analysis Tools to Test a Claim About Two Means When Standard Deviations Are Unknown and Assumed to Be Equal

Exercise 10 asks you to use a 0.01 significance level to test the claim that men and women have different body temperatures. We will assume that population standard deviations are unknown and are assumed to be equal. You will carry out a *t*-test for two independent samples using Excel's Data Analysis Tools.

1. The instructions in Exercise 10 tell you to use the body temperatures listed in the last column of Data Set 3 in Appendix B. Rather than use the entire data set, we will use just the first 12 values for the males and females. Enter the males' and females' body temperatures in an Excel worksheet as shown below.

	A	B
1	Male	Female
2	98.6	97.7
3	98.6	98.8
4	98	98
5	98	98.7
6	99	98.4
7	98.4	98
8	98.4	97.9
9	98.4	98.2
10	98.4	98
11	98.6	98.5
12	98.6	98.3
13	98.8	98.7

2. Click **DATA** at the top of the screen and select **Data Analysis** on the far right of the data ribbon.

If Data Analysis does not appear as a choice in the data ribbon, you will need to load the Microsoft Excel ToolPak add-in. Follow the procedure on page 10.

3. The Data Analysis dialog box presents three *t*-test options. For this problem, you want the second one, ***t*-Test: Two-Sample Assuming Equal Variances**. Click **OK**.

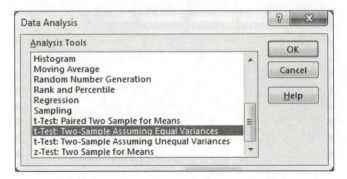

4. Complete the dialog box as shown below. A description of the entries immediately follows the dialog box.

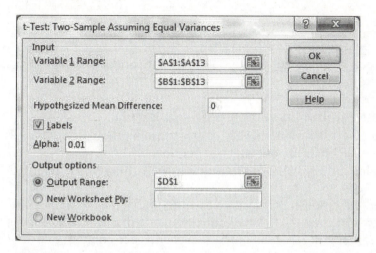

- **Variable 1 Range**: Enter the worksheet location of the males' data, **A1:A13**.
- **Variable 2 Range**: Enter the worksheet location of the females' data, **B1:B13**.
- **Hypothesized Mean Difference**: The default value is zero. It is not necessary to enter anything here if you are using the default value.
- **Labels**: A check mark should appear in the box to the left of Labels to indicate that the top cell in each variable range is a label (Male or Female) and not a data value.
- **Alpha**: The instructions in Exercise 10 ask you to use a 0.01 significance level, so enter **0.01** in the Alpha window.
- **Output options**: You are given three options for placement of the *t*-test output. Let's place the output in the same worksheet as the data. The top leftmost cell of the output will be **D1**.

5. Click **OK**.

The Excel output is displayed below. You will want to make column D wider so that you can see all the output labels. An explanation of the output is presented on the next page.

	A	B	C	D	E	F
1	Male	Female		t-Test: Two-Sample Assuming Equal Variances		
2	98.6	97.7				
3	98.6	98.8			Male	Female
4	98	98		Mean	98.48333	98.26667
5	98	98.7		Variance	0.083333	0.127879
6	99	98.4		Observations	12	12
7	98.4	98		Pooled Variance	0.105606	
8	98.4	97.9		Hypothesized Mean Difference	0	
9	98.4	98.2		df	22	
10	98.4	98		t Stat	1.63314	
11	98.6	98.5		P(T<=t) one-tail	0.058336	
12	98.6	98.3		t Critical one-tail	2.508325	
13	98.8	98.7		P(T<=t) two-tail	0.116671	
14				t Critical two-tail	2.818756	

- **Mean**: The sample mean for males is 98.48 and the sample mean for females is 98.27.
- **Variance**: The variance for the males' data is 0.083 and the variance for the females' data is 0.128.
- **Observations**: The sample size for males is 12 and the sample size for females is 12.
- **Pooled variance**: The pooled variance for the *t*-test is 0.106.
- **Hypothesized mean difference**: The mean difference stated in the null hypothesis is 0.
- **df**: Degrees of freedom for the test equal 22.
- **t stat**: The observed value of the *t* statistic is 1.633.
- **P(T<=t) one-tail**: The one-tailed probability of observed *t* equals 0.058.
- **t critical one-tail**: The one-tailed critical value of *t* for your chosen alpha (0.01) equals 2.508.
- **P(T<=t) two-tail**: The two-tailed probability of observed *t* equals 0.117.
- **t critical two-tail**: The two-tailed critical value of *t* for your chosen alpha (0.01) equals 2.819.

Test Interpretation: If you carried out a two-tailed test, you would not reject the null hypothesis because the two-tailed *p*-value is greater than 0.01. You would conclude that the difference is not statistically significant and that the population means of the two groups are equal to one another.

Exercise 10, page 464	**Using XLSTAT to Test a Claim About Two Means When Standard Deviations Are Unknown and Assumed to Be Equal**

To explain how to use XLSTAT for the situation involving unknown population standard deviations that are assumed to be equal, we will use the same Exercise 10 data that we used for illustrating Data Analysis Tools.

1. If you have not already done so, enter the male and female body temperature data in an Excel worksheet. These data are shown on page 102 of this manual.
2. At the top of the screen, select the **XLSTAT** add-in.
3. Select **Parametric tests**. Select **Two-sample t test and z test**.
4. Complete the General dialog box as shown below. A description of the entries is given immediately after the dialog box.

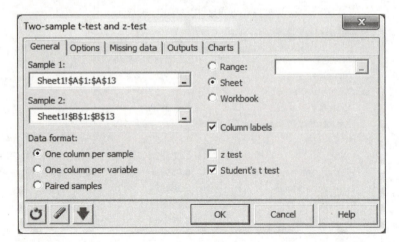

- **Sample 1**: Enter the range of the males' data, **A1:A13**.

- **Sample 2**: Enter the range of the females' data, **B1:B13**.
- **Data format**: Select **One column per sample**.
- **Sheet**: The output will be placed in a new Excel worksheet.
- **Column labels**: Because the top cells in the data ranges are labels and not data values, you need to select **Column labels**.
- **Student's t test**: Because the population standard deviations are not known and are estimated from the sample data, the appropriate test is the **Student's *t*-test**.

5. Click the **Options** tab at the top of the dialog box. Complete the Options dialog box as shown below. A description of the entries is given immediately after the dialog box.

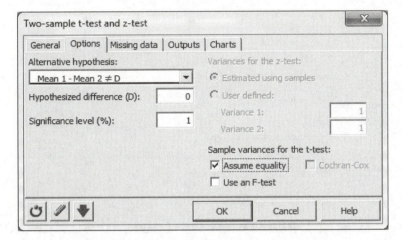

- **Alternative hypothesis**: We will be carrying out a two-tailed test where the alternative hypothesis is **Mean 1 – Mean 2 ≠ D**.
- **Hypothesized difference (D)**: The hypothesized mean difference is **0**.
- **Significance level**: The significance level is **1** (or 1%).
- **Sample variances for the t-test**: For the variances for the *t*-test, select **Assume equality**.

6. Click the **Outputs** tab at the top of the dialog box.   Be sure there is a check mark in the box next to **Descriptive Statistics**.
7. Click **OK**. Click **Continue** in the XLSTAT-Selections dialog box.

The XLSTAT output is shown below and at the top of the next page. The descriptive statistics are displayed first.

Summary statistics:							
Variable	Observations	Obs. with missing data	Obs. without missing data	Minimum	Maximum	Mean	Std. deviation
Male	12	0	12	98.0000	99.0000	98.4833	0.2887
Female	12	0	12	97.7000	98.8000	98.2667	0.3576

The next part of the XLSTAT output presents the 99% two-tailed confidence interval and the results of the *t*-test for two independent samples. The test interpretation at the bottom of the XLSTAT output indicates that the result is not statistically significant.

t-test for two independent samples / Two-tailed test:			
99% confidence interval on the difference between the means:			
(-0.1573 , 0.5906)			
Difference	0.2167		
t (Observed value)	1.6331		
|t| (Critical value)	2.8188		
DF	22		
p-value (Two-tailed)	0.1167		
alpha	0.01		
Test interpretation:			
H0: The difference between the means is equal to 0.			
Ha: The difference between the means is different from 0.			
As the computed p-value is greater than the significance level alpha=0.01, one cannot reject the null hypothesis H0.			
The risk to reject the null hypothesis H0 while it is true is 11.67%.			

Exercise 16, page 465	# Using Data Analysis Tools to Test a Claim About Two Means When Standard Deviations Are Known

To illustrate how to use Excel's Data Analysis Tools to carry out a *z*-test to test a claim regarding the equality of two population means, we will use the full-scale IQ scores for low and high lead level groups in the IQLEAD data file. Exercise 16 on page 465 of the Triola text asks you to test the claim that the mean IQ score of people with low lead levels is higher than the mean IQ score of people with high lead levels. The Wechsler Adult Intelligence Scale (WAIS) is often used to test the IQ of adult subjects, and the population standard deviation of the WAIS is equal to 15. For this problem, we will assume that participants took the WAIS, and we will assume that the population standard deviation is known and is equal to 15.

1. Begin by opening the **IQLEAD** data file on your data disk.
2. Column A of the IQLEAD data file contains the lead level condition (LEAD). Low, medium, and high lead levels are coded as 1, 2, and 3, respectively.   Column H contains the full-scale IQ (IQF).   The first few rows are shown below.

	A	B	C	D	E	F	G	H
1	LEAD	AGE	SEX	YEAR1	YEAR2	IQV	IQP	IQF
2	1	11	1	25	18	61	85	70
3	1	9	1	31	28	82	90	85
4	1	11	1	30	29	70	107	86
5	1	6	1	29	30	72	85	76

To carry out a z-test using Excel's Data Analysis Tools, you will need to place the full-scale IQ's for condition 1 (low lead level) in one column of an Excel worksheet, and the full-scale IQ's for condition 3 (high lead level) in another. Open a new Excel worksheet and type **Low** in cell **A1** and **High** in cell **B1**. Then place the condition 1 IQF data in column A and the condition 3 IQF data in column B. The first few rows are shown below.

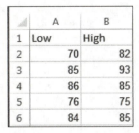

	A	B
1	Low	High
2	70	82
3	85	93
4	86	85
5	76	75
6	84	85

3. Click **DATA** at the top of the screen and select **Data Analysis** on the far right of the data ribbon.

If Data Analysis does not appear as a choice in the data ribbon, you will need to load the Microsoft Excel ToolPak add-in. Follow the procedure on page 10.

4. Select **z-Test: Two Sample for Means**. Click **OK**.

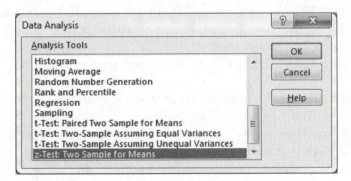

5. Complete dialog box as shown at the top of the next page. A description of the entries is given immediately after the dialog box.

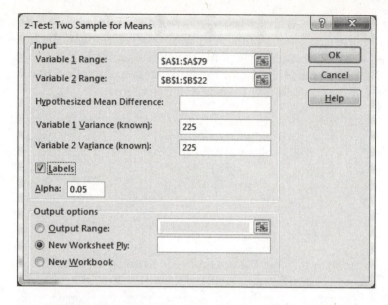

- **Variable 1 Range**: Enter the range of the low lead group's full-scale IQ's, **A1:A79**.
- **Variable 2 Range**: Enter the range of the high lead group's full-scale IQ's, **B1:B22**.
- **Hypothesized Mean Difference**: The default value is zero. It is not necessary to enter anything here if you are using the default value.
- **Variable 1 Variance (known)**: The variance is equal to the standard deviation squared. For this exercise, we are assuming that the population standard deviation is equal to 15. Because $15^2 = 225$, we enter **225** for the variance.
- **Variable 2 Variance (known)**: Enter **225**, the same value you entered for the variable 1 variance.
- **Labels**. A check mark in the box to the left of **Labels** indicates that the top cell in each variable range is a label (Low or High) and not a data value.
- **Alpha**. We want to set alpha equal to 0.05 for this test, and because **0.05** is the default value, we do not need to change the value.
- **Output options**. Select **New Worksheet Ply** for placement of the output.

6. Click **OK**. The output is shown below. You will want to make column A wide enough so that you can see all the output labels. An explanation of entries is given immediately after the output.

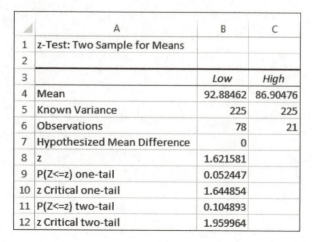

	A	B	C
1	z-Test: Two Sample for Means		
2			
3		Low	High
4	Mean	92.88462	86.90476
5	Known Variance	225	225
6	Observations	78	21
7	Hypothesized Mean Difference	0	
8	z	1.621581	
9	P(Z<=z) one-tail	0.052447	
10	z Critical one-tail	1.644854	
11	P(Z<=z) two-tail	0.104893	
12	z Critical two-tail	1.959964	

- **Mean**: The sample mean for low lead group is 92.88 and the sample mean for the high lead group is 86.90.
- **Known variance**: The known variance for both groups is equal to 225.
- **Observations**: The sample size for the low lead group is 78 and the sample size for the high lead group is 21.
- **Hypothesized mean difference**: The mean difference stated in the null hypothesis is 0.
- **z**: The observed value of the z statistic is 1.622.
- **P(Z<=z) one-tail**: The one-tailed probability of observed z equals 0.052.
- **z critical one-tail**: The one-tailed critical value of z for your chosen alpha (0.05) equals 1.645.
- **P(Z<=z) two-tail**: The two-tailed probability of observed z equals 0.105.
- **z critical two-tail**: The two-tailed critical value of z for your chosen alpha (0.05) equals 1.960.

Test Interpretation: If you carried out a one-tailed test with an alternative hypothesis stating that the low lead group's mean is greater than the high lead group's mean, you would conclude that the difference is not statistically significant because the one-tailed p-value (0.052) is greater than 0.05. You would conclude that the population means of the two groups are equal to one another.

Exercise 16, page 465	# Using XLSTAT to Test a Claim About Two Means When Standard Deviations Are Known

To illustrate how to use XLSTAT to carry out a z-test to test a claim regarding the equality of two population means, we will use the same IQ data that we just used to illustrate the use of Excel's Data Analysis Tools. Exercise 16 asks you to test the claim that the mean IQ score of people with low lead levels is higher than the mean IQ score of people with high lead levels. Recall that we are assuming that the population standard deviation of the IQ scores is known and that it is equal to 15 for both groups.

1. If you have not already do so, complete steps 1 and 2 on page 106 of this manual. The first few rows of the data are shown below.

◢	A	B
1	Low	High
2	70	82
3	85	93
4	86	85
5	76	75
6	84	85

2. At the top of the screen, select the **XLSTAT** add-in.
3. Select **Parametric tests**. Select **Two-sample t test and z test**.
4. Complete the General dialog box as shown at the top of the next page. Descriptions of the entries are provided immediately after the dialog box.

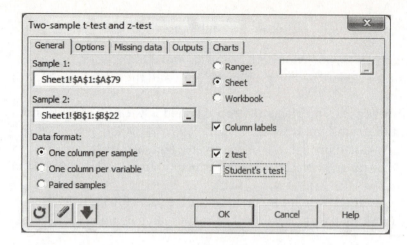

- **Sample 1**: Enter the range of IQF data for the low lead group, **A1:A79**.
- **Sample 2**: Enter the IQF range for the high lead group, **B1:B22**.
- **Sheet**: The output will be placed in a new Excel worksheet.
- **Column labels**: Because the top cells in the data ranges are labels and not data values, you need to select **Column labels**.
- **Test**: Because the population variances are known, the appropriate test is the **z-test**.

5. Click the **Options** tab at the top of the dialog box. Complete the dialog box as shown below. A description of the entries is given immediately after the dialog box.

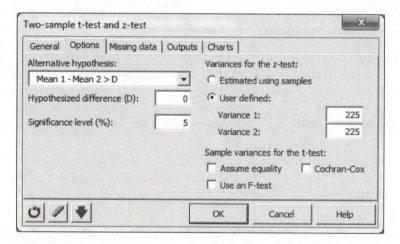

- **Alternative hypothesis**: Select the one-tailed alternative hypothesis that indicates that the mean of the low lead group is greater than the mean of the high lead group, **Mean 1 – Mean 2 > D**.
- **Hypothesized difference (D)**: The hypothesized difference is **0**.
- **Significance level**: The significance level is **5** (or 5%).
- **Variances for the z-test**: Select **User defined** variances for the z-test. Enter **225** for both variance 1 and variance 2.

6. Click the **Outputs** tab at the top of the dialog box. Be sure that there is a check mark in the box next to **Descriptive statistics**.
7. Click **OK**. Click **Continue** in the XLSTAT-Selections dialog box.

The first section of the XLSTAT output contains summary statistics. The summary statistics include the number of observations in each group, the minimum, the maximum, the mean and the standard deviation.

Summary statistics:							
Variable	Observations	Obs. with missing data	Obs. without missing data	Minimum	Maximum	Mean	Std. deviation
Low	78	0	78	50.0000	141.0000	92.8846	15.3445
High	21	0	21	75.0000	104.0000	86.9048	8.9884

The next section of the XLSTAT output is labeled *z-test for two independent samples/Upper-tailed test*. The output includes the one-tailed 95% confidence interval, the results of the z-test, and the test interpretation.

z-test for two independent samples / Upper-tailed test:	
95% confidence interval on the difference between the means:	
(-0.0858 , +Inf)	
Difference	5.9799
z (Observed value)	1.6216
z (Critical value)	1.6449
p-value (one-tailed)	0.0524
alpha	0.05
Test interpretation:	
H0: The difference between the means is equal to 0.	
Ha: The difference between the means is greater than 0.	
As the computed p-value is greater than the significance level alpha=0.05, one cannot reject the null hypothesis H0.	
The risk to reject the null hypothesis H0 while it is true is 5.24%.	

9-4 Matched Pairs

Exercise 5, page 474	**Using Data Analysis Tools to Test a Claim About the Means of Paired Samples**

Exercise 5 asks you to use all the cases in Data Set 12 in Appendix B to test the claim that, for the population of heights of presidents and their main opponents, the differences have a mean greater than 0 cm (so that presidents tend to be taller than their opponents). You are asked to use a 0.05 significance level. The data are located in the POTUS data file on your data disk.

1. Begin by opening the **POTUS** data file on your data disk.
2. The heights of the presidents are located in column E and the variable label is Ht. The heights of the opponents are located in column F and the variable label is HtOpp. Some of the opponents' heights are missing. In order to use Data Analysis Tools for this analysis, you

can include only the data where both pair members are present. Copy the Ht and HtOpp data to sheet 2 and delete the rows that have missing data. The first few rows of the revised data set are shown below.

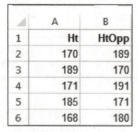

	A	B
1	Ht	HtOpp
2	170	189
3	189	170
4	171	191
5	185	171
6	168	180

3. Click **DATA** at the top of the screen and select **Data Analysis** on the far right of the data ribbon.

If Data Analysis does not appear as a choice in the data ribbon, you will need to load the Microsoft Excel ToolPak add-in. Follow the procedure on page 10.

4. Select **t-Test: Paired Two Sample for Means**. Click **OK**.

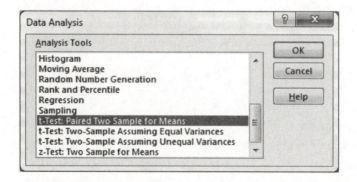

5. Complete the dialog box as shown below. A description of the entries is given immediately after the dialog box.

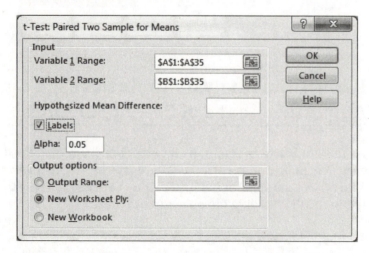

- **Variable 1 Range**: Enter the worksheet location of the presidents' heights, **A1:A35**.
- **Variable 2 Range**: Enter the worksheet location of the opponents' heights, **B1:B35**.

- **Hypothesized Mean Difference**: The default value is zero. It is not necessary to enter anything here if you are using the default value.
- **Labels**: A check mark in the box to the left of **Labels** indicates that the top cell in each range is a label (Ht or HtOpp) and not a data value.
- **Alpha**: The instructions state that you should use a significance level of **0.05**.
- **Output Options**: Let's place the output in a **New Worksheet Ply**.

6. Click **OK**. The output is shown below. An explanation immediately follows the output.

	A	B	C
1	t-Test: Paired Two Sample for Means		
2			
3		Ht	HtOpp
4	Mean	180.0294	179.9706
5	Variance	47.54456	38.45365
6	Observations	34	34
7	Pearson Correlation	-0.07298	
8	Hypothesized Mean Difference	0	
9	df	33	
10	t Stat	0.035714	
11	P(T<=t) one-tail	0.485863	
12	t Critical one-tail	1.69236	
13	P(T<=t) two-tail	0.971726	
14	t Critical two-tail	2.034515	

- **Mean**: The mean of the presidents' heights is 180.03 cm and the mean of their opponents' heights is 179.97 cm.
- **Variance**: The variance of the presidents' heights is 47.54 and the variance of their opponents' heights is 38.45.
- **Observations**: Both samples have 34 observations.
- **Pearson Correlation**: The Pearson correlation between the heights of the presidents and their opponents is -0.073.
- **Hypothesized Mean Difference**: The hypothesized mean difference is 0.
- **df**: The degrees of freedom for the test equal 33.
- **t Stat**: The observed t statistic is 0.036.
- **P(T<=t) one-tail**: The one-tailed p-value of observed t is equal to 0.486.
- **t Critical one-tail**: The one-tailed critical value of t with 33 degrees of freedom and a significance level of 0.05 is 1.692.
- **P(T<=t) two-tail**: The two-tailed p-value of observed t is equal to 0.972.
- **t Critical two-tail**: The two-tailed critical value of t with 33 degrees of freedom and a significance level of 0.05 is 2.035.

Test Interpretation: If you were doing a one-tailed test, you would conclude that the result is not statistically significant because the one-tailed p-value (0.486) is greater than 0.05. If you were doing a two-tailed test, you would also conclude that the result is not statistically significant because the two-tailed p-value (0.972) is greater than 0.05. Based on the results for either the one-tailed or two-tailed test, you would conclude that the average difference is equal to 0.

Exercise 5, page 474	# Using XLSTAT to Test a Claim About the Means of Paired Samples

To illustrate how to use XLSTAT to carry out a paired-samples *t*-test, we will use the heights of presidents and their opponents that are found on the POTUS data file on your data disk. To use Excel's Data Analysis Tools for this analysis, you first had to revise the data set so that it did not contain any missing values. This will not be necessary when you use XLSTAT, because XLSTAT allows you to select a procedure for dealing with missing data.

1. Begin by opening the **POTUS** data file on your data disk.
2. At the top of the screen, select the **XLSTAT** add-in.
3. Select **Parametric tests**. Select **Two-sample t test and z test**.
4. Complete the General dialog box as shown below. A description of the entries is given immediately after the dialog box.

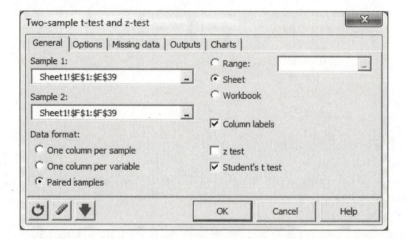

- **Sample 1**: Enter the range for the Ht variable (the heights of presidents), **E1:E39**.
- **Sample 2**: Enter the range for the HtOpp variable (the heights of the opponents), **F1:F39**.
- **Data format**: Select **Paired samples**.
- **Sheet**: The output will be placed in a new Excel worksheet.
- **Column labels**: Be sure to check **Column labels** because the top cell in each data range is label, not a data value.
- **Test**: Select the **Student's *t* test**.

5. Click the **Options** tab at the top of the dialog box. Complete the dialog box as shown at the top of the next page. A description of the entries is given immediately after the dialog box.

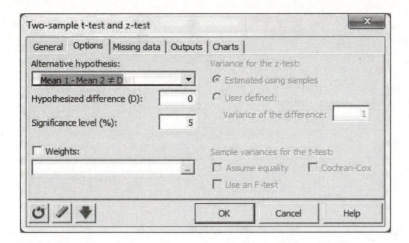

- **Alternative hypothesis**: Select **Mean 1 – Mean 2 ≠ D** to carry out a two-tailed test.
- **Hypothesized difference (D):** The hypothesized difference is equal to **0**.
- **Significance level:** The significance level of the test is **5%**.

6. Click the **Missing data** tab at the top of the dialog box. Select **Remove the observations**.

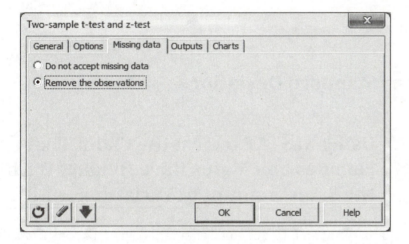

7. Click **OK**. Click **Continue** in the XLSTAT-Selections dialog box.

The XLSTAT output includes summary statistics and the results of the *t*-test for two paired samples. The summary statistics are shown below.

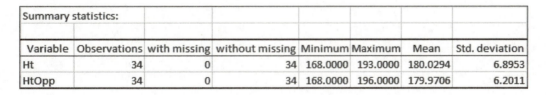

Summary statistics:							
Variable	Observations	with missing	without missing	Minimum	Maximum	Mean	Std. deviation
Ht	34	0	34	168.0000	193.0000	180.0294	6.8953
HtOpp	34	0	34	168.0000	196.0000	179.9706	6.2011

The XLSTAT output for the *t-test for two paired samples* includes a confidence interval, the *t*-test results, and a test interpretation.

t-test for two paired samples / Two-tailed test:					
95% confidence interval on the difference between the means:					
(-3.2922 ,	3.4099)				
Difference	0.0588				
t (Observed value)	0.0357				
|t| (Critical value)	2.0345				
DF	33				
p-value (Two-tailed)	0.9717				
alpha	0.05				
Test interpretation:					
H0: The difference between the means is equal to 0.					
Ha: The difference between the means is different from 0.					
As the computed p-value is greater than the significance level alpha=0.05, one cannot reject the null hypothesis H0.					
The risk to reject the null hypothesis H0 while it is true is 97.17%.					

9-5   Two Variances or Standard Deviations

Exercise 15, page 485	**Using XLSTAT to Test the Claim That Females and Males Have Heights With the Same Amount of Variation**

To illustrate how to use XLSTAT to carry out an *F*-test of equal standard deviations (or variances), we will be using the data referenced in Exercise 15 on page 485 of the Triola text. Exercise 15 asks you to test the claim that females and males have heights with the same amount of variation.

1. Begin by entering the Exercise 15 data in an Excel worksheet as shown at the top of the next page.

	A	B
1	**F-Height**	**M-Height**
2	163.7	178.8
3	165.5	177.5
4	163.1	187.8
5	166.3	172.4
6	163.6	181.7
7	170.9	169.0
8	153.5	186.9
9	155.7	183.1
10	153.0	176.4
11	157.0	183.4

2. For this *F*-test, the sample with the larger variance needs to be entered as sample 1 and the sample with the smaller variance as sample 2. To find out which sample has the larger variance, you will use XLSTAT to obtain descriptive statistics for the two samples. At the top of the screen, select the **XLSTAT** add-in.

3. Select **Describing data**. Select **Descriptive statistics**.

4. Complete the General dialog box as shown below. A description of the entries is given immediately after the dialog box.

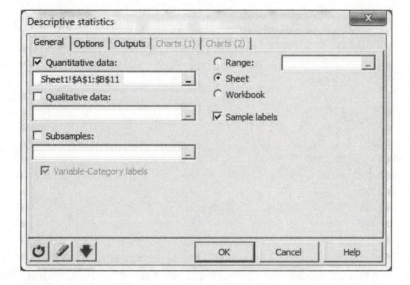

- **Quantitative data**: Select **Quantitative data** and enter the worksheet range of both samples, **A1:B11**.
- **Sheet**: The output will be placed in a new Excel worksheet.
- **Sample labels**: A check mark here tells XLSTAT that the top cells are column labels and not data values.

5. Click the **Options** tab. Make sure that **Descriptive statistics** has been selected.

6. Click the **Outputs tab**. You will want to select **Variance (n-1)** and/or **Standard deviation (n-1)**. I have selected both of these options.

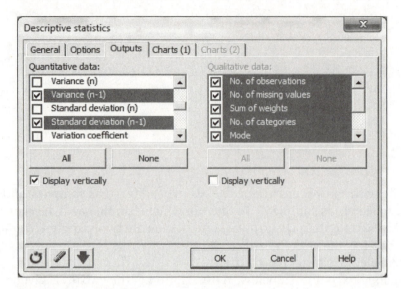

7. Click **OK**. Click **Continue** in the XLSTAT-Selections dialog box.
8. The XLSTAT descriptive statistics output is shown below. The variance of the males' heights (36.7800) is larger than the variance of the females' heights (36.5134). Therefore, the males' data will need to be entered as sample 1 and the females' data as sample 2.

Statistic	F-Height	M-Height
No. of observations	10	10
Minimum	153.0000	169.0000
Maximum	170.9000	187.8000
1st Quartile	156.0250	176.6750
Median	163.3500	180.2500
3rd Quartile	165.0500	183.3250
Mean	161.2300	179.7000
Variance (n-1)	36.5134	36.7800
Standard deviation (n-1)	6.0426	6.0647

9. Select **Parametric tests**. Select **Two-sample comparison of variances**.
10. Complete the General dialog box as shown at the top of the next page. A description of the entries is given immediately after the dialog box.

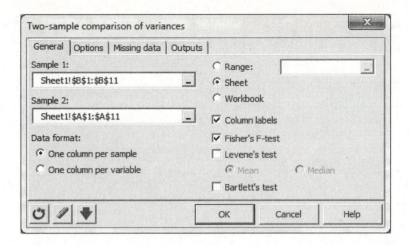

- **Sample 1**: Enter the range of the males' height data, **B1:B11**. Recall that the sample with the larger variance needs to be sample 1.
- **Sample 2**: Enter the range of the females' height data, **A1:A11**.
- **Data format**: Select **One column per sample**.
- **Sheet**: The output will be placed in a new Excel worksheet.
- **Column labels**. Select **Column labels** so that XLSTAT will treat the top cell in each range as a label and not a data value.
- **Fishers' F-test**. Select **Fisher's F-test** to carry out an F-test of equal variances.

11. Click the **Options** tab. Select the two-tailed alternative hypothesis, **Variance 1/Variance 2 ≠ R**. Set the significance level equal to **5**%.

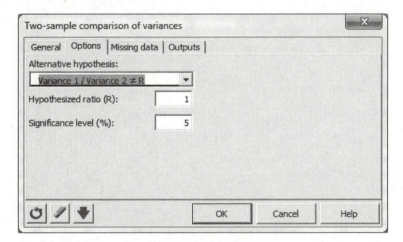

12. There are no missing data, so you don't need to select any missing data options. Click **Outputs** and select **Descriptive statistics**.
13. Click **OK**. Click **Continue** in the XLSTAT-Selections dialog box.

The XLSTAT output for the F-test is shown below. The observed F is equal to 1.0073, and the two-tailed p-value of observed F is 0.9915. Because the p-value is greater than 0.05, the result is not statistically significant.

B	C	D	E	F
Fisher's F-test / Two-tailed test:				
95% confidence interval on the ratio of variances:				
(0.2502,	4.0554)			
Ratio	1.0073			
F (Observed value)	1.0073			
F (Critical value)	4.0260			
DF1	9			
DF2	9			
p-value (Two-tailed)	0.9915			
alpha	0.05			
Test interpretation:				
H0: The ratio between the variances is equal to 1.				
Ha: The ratio between the variances is different from 1.				
As the computed p-value is greater than the significance level alpha=0.05, one cannot reject the null hypothesis H0.				
The risk to reject the null hypothesis H0 while it is true is 99.15%.				

10

Correlation and Regression

10-2 Correlation

Using XLSTAT to Test for a Relationship Between Length of Shoe Print and Height

You will be using XLSTAT to test for a significant linear relationship between two quantitative variables assessed by the Pearson correlation coefficient. The XLSTAT output provides summary descriptive statistics, a correlation matrix, the coefficient of determination, the *p*-value associated with the *t*-test, and a scatterplot. To illustrate how to use XLSTAT for this analysis, we will use the data referenced in Example 5 on page 503 of the Triola text. For this analysis, the *x* variable is length of shoe print and the *y* variable is height. You are addressing the question of whether or not there is sufficient evidence to support a claim of a linear correlation between the lengths of shoe prints and heights.

1. Begin by opening the **FOOT** data file on your data disk.
2. The *x* and *y* variables to be included in the analysis need to be located in adjacent columns of an Excel worksheet. I copied the sheet 1 columns containing Shoe Print and Height and pasted them in sheet 2. The first few rows are shown below.

	A	B
1	**Shoe Print**	**Height**
2	31.3	180.3
3	29.7	175.3
4	31.3	184.8
5	31.8	177.8
6	31.4	182.3

3. At the top of the screen, select the **XLSTAT** add-in.
4. Select **Correlation/Association tests**. Select **Correlation tests**.
5. Complete the General dialog box as shown below. A description of the entries is given immediately after the dialog box.

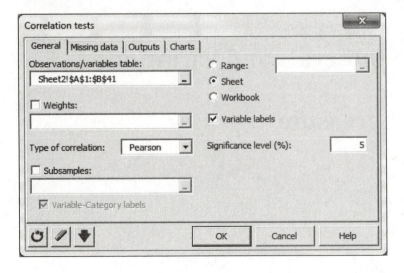

- **Observations/variables table**: Enter the worksheet range that contains the shoe print and height data, **A1:B41**.
- **Type of correlation**: XLSTAT provides three options: Pearson, Spearman, and Kendall. Select **Pearson**.
- **Sheet**: The output will be placed in a new Excel worksheet.
- **Variable labels**: Be sure to place a check mark in the box next to **Variable labels**. This check mark lets XLSTAT know that the top cell in each column is a label and not a data value.
- **Significance level (%)**: Set the significance level at **5** (or 5%) for the *t*-test.

6. Click the **Outputs** tab. Select all the output options as shown below.

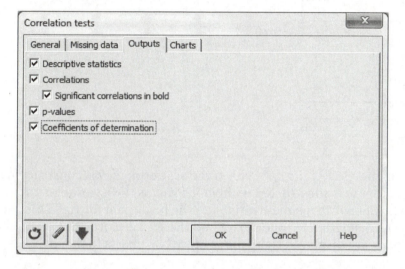

7. Click the **Charts** tab. Select **Scatter plots** and **Histograms**. The histograms output gives you information about the frequencies associated with the *x* and *y* variables.
8. Click **OK**. Click **Continue** in the XLSTAT-Selections dialog box.

The first part of the XLSTAT output presents descriptive statistics for the Shoe Print and Height variables. The second part presents the correlation matrix. The Pearson correlation coefficient (*r*) is equal to 0.813.

Summary statistics:							
Variable	Observations	Obs. with missing data	Obs. without missing data	Minimum	Maximum	Mean	Std. deviation
Shoe Print	40	0	40	24.8000	34.5000	29.0175	2.5448
Height	40	0	40	152.4000	195.0000	174.3250	10.0750

Correlation matrix (Pearson):		
Variables	Shoe Print	Height
Shoe Print	1	0.8129
Height	0.8129	1
Values in bold are different from 0 with a significance level alpha=0.05		

The XLSTAT output also presents the *p*-value associated with the obtained *t*-test result. The *p*-value for this analysis is very small and is shown in the output as < 0.0001. In addition, the *p*-value is shown in bold type. The note below the *p*-value tells you that the bold indicates that the value is different from 0 with a significance level (alpha) equal to 0.05. Note that the *t*-test itself is not displayed in the output. The coefficient of determination output follows the *p*-value output. The coefficient of determination is equal to 0.661.

p-values:			
Variables	Shoe Print	Height	
Shoe Print	0	< 0.0001	
Height	< 0.0001	0	
Values in bold are different from 0 with a significance level alpha=0.05			
Coefficients of determination (R²):			
Variables	Shoe Print	Height	
Shoe Print	1	0.6609	
Height	0.6609	1	

Two scatterplots are presented in the XLSTAT output. One of the scatterplots was constructed with Height as the *x* variable and the other was constructed with Shoe Print as the x variable. You want the one that was constructed with Shoe Print as the *x* variable.   The histogram for Height is shown to the right of this scatterplot. The histogram for Shoe Print and the scatterplot with Height as the *x* variable, which are not shown here, are displayed right above these two charts in the XLSTAT output.

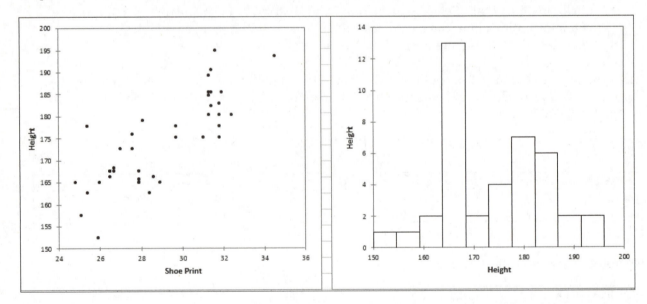

10-3 Regression

Example 4, page 521	Using XLSTAT to Carry Out a Regression Analysis on Length of Shoe Print and Height

The output from XLSTAT's linear regression procedure includes descriptive statistics for the *x* and *y* variables, a correlation matrix, goodness of fit statistics, analysis of variance, the prediction equation, the standardized regression coefficient, predicted *y* values, and residuals. To illustrate how to use XLSTAT to carry out a regression analysis, we will use the data referenced in part b of Example 4 on page 521 of the Triola text. For this analysis, you will use the 40 pairs of shoe print lengths and heights in the FOOT data file to calculate the prediction equation to predict height from shoe print length.

1. Begin by opening the **FOOT** data file on your data disk. The data you will be working with are in sheet 1. Shoe Print is in column D and Height is in column F
2. At the top of the screen, select the **XLSTAT** add-in.
3. Select **Modeling data**. Select **Linear regression**.
4. Complete the General dialog box as shown below. A description of the entries is given immediately after the dialog box.

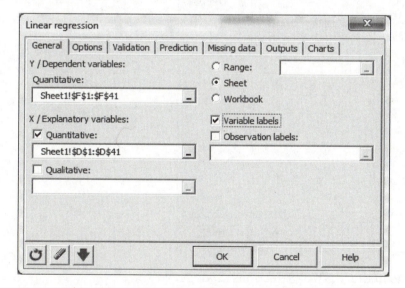

- **Y/Dependent variables: Quantitative**: Enter the range of the Height data, **F1:F41**.
- **X/Explanatory variables: Quantitative**: Enter the range of the Shoe Print data, **D1:D41**.
- **Sheet**: The output will be placed in a new Excel worksheet.
- **Variable labels**: Be sure to place a check mark in the box next to **Variable labels**. This check mark lets XLSTAT know that the top cells in the data ranges are labels and not data values.

5. Click the **Options** tab. We will use the default value for the confidence interval, 95%. So, we don't need to enter or select anything here.

6. You can skip the Validation and Prediction tabs. Because there are no missing data in the FOOT data file, you can also skip the Missing Data tab.

7. Click the **Outputs** tab. We want same five outputs that are already selected.

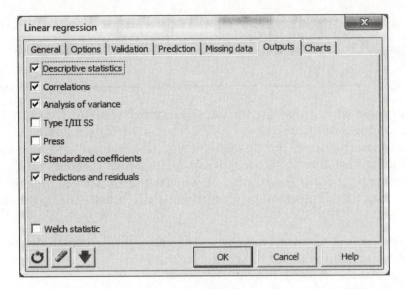

8. Click the **Charts** tab. We want all the chart outputs, and these have already been selected.

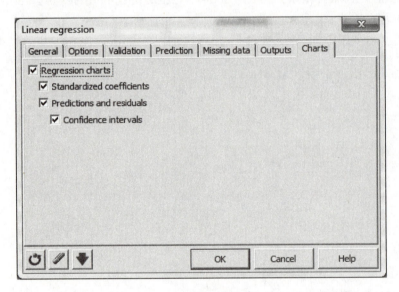

9. Click **OK**. Click **Continue** in the XLSTAT-Selections dialog box.

The XLSTAT output is shown below. Summary statistics for the Height and Shoe Print variables are shown first followed by the correlation matrix. The Pearson correlation, r, between Shoe Print and Height is equal to 0.813.

Summary statistics:							
Variable	Observations	Obs. with missing data	Obs. without missing data	Minimum	Maximum	Mean	Std. deviation
Height	40	0	40	152.4000	195.0000	174.3250	10.0750
Shoe Print	40	0	40	24.8000	34.5000	29.0175	2.5448

Correlation matrix:		
Variables	Shoe Print	Height
Shoe Print	**1.0000**	0.8129
Height	0.8129	**1.0000**

The goodness of fit statistics provided in the XLSTAT output include the number of observations, the sum of the weights, degrees of freedom, R^2, adjusted R^2, mean square error (MSE), root mean square error (RMSE), and the Durbin–Watson statistic (DW). The analysis of variance (ANOVA) output displays the results of an F test that is carried out to test whether or not the slope of the regression line is significantly different from zero. The p-value shown in the ANOVA output (Pr >F) is < 0.0001, indicating the result is statistically significant and the slope is significantly different from zero.

Regression of variable Height:	
Goodness of fit statistics:	
Observations	40.0000
Sum of weights	40.0000
DF	38.0000
R²	0.6609
Adjusted R²	0.6520
MSE	35.3283
RMSE	5.9438
DW	1.7296

Analysis of variance:					
Source	DF	Sum of squares	Mean squares	F	Pr > F
Model	1	2616.2796	2616.2796	74.0562	< 0.0001
Error	38	1342.4754	35.3283		
Corrected Total	39	3958.7550			
Computed against model Y=Mean(Y)					

XLSTAT's model parameters output includes the intercept (80.930) and the slope (3.219) of the regression equation, significance tests associated the intercept and slope, and lower and upper bounds of the 95% confidence intervals formed around the intercept and slope. The prediction equation is shown next. You can use this equation to answer part b of Example 4 where you are asked to predict the height of a person with a shoe print length of 29 cm. The standardized coefficient output displays the standardized value of the slope, 0.813. In the case of bivariate regression which has one predictor variable, the standardized regression coefficient is equal to r.

Model parameters:						
Source	Value	Standard error	t	Pr > |t|	Lower bound (95%)	Upper bound (95%)
Intercept	80.9304	10.8934	7.4293	< 0.0001	58.8779	102.9829
Shoe Print	3.2186	0.3740	8.6056	< 0.0001	2.4614	3.9757
Equation of the model:						
Height = 80.93041+3.21856*Shoe Print						
Standardized coefficients:						
Source	Value	Standard error	t	Pr > |t|	Lower bound (95%)	Upper bound (95%)
Shoe Print	0.8129	0.0945	8.6056	< 0.0001	0.6217	1.0042

Additional output, not shown here, presents information related to predictions and residuals, such observed shoe print lengths, observed heights, predicted heights, residuals, and standardized residuals. Charts that display the regression line and standardized residuals are also included in the XLSTAT output.

10-4   Prediction Intervals and Variation

Example 1, pages 533-536	**Using XLSTAT to Carry Out a Regression Analysis on Length of Shoe Print and Height**

In Section 10-3, you carried out a regression analysis on the FOOT data to obtain the equation for predicting height from shoe print length. Example 1 in Section 10-4 on page 533 of the Triola text asks you to predict height for a shoe print length equal to 29 cm and then to form a 95% prediction interval around the predicted height. Example 2 on page 536 asks you to find the coefficient of determination. To illustrate how to use XLSTAT to carry out these calculations, we will use the FOOT data file located on your data disk.

1. Begin by opening the **FOOT** data file on your data disk. The data will you will be analyzing are in sheet 1. Shoe Print is in column D and Height is in column F.
2. To find the predicted height for a shoe print length of 29 cm, you need to enter 29 somewhere on the worksheet. I added the label **New X** in cell **G1** and typed **29** in cell **G2** as shown below. The heading (New X) isn't necessary, but it might serve to identify where you have placed the *x* values for which you want to predict *y*'s.

	A	B	C	D	E	F	G
1	Sex	Age	Foot Length	Shoe Print	Shoe Size	Height	New X
2	M	67	27.8	31.3	11	180.3	29.0
3	M	47	25.7	29.7	9	175.3	

3. At the top of the screen, select the **XLSTAT** add-in.
4. Select **Modeling data**. Select **Linear regression**.
5. Complete the General dialog box as shown below. A description of the entries is given immediately after the dialog box.

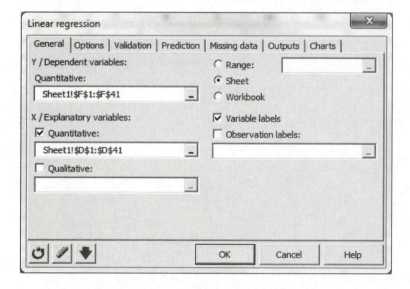

- **Y/Dependent variables: Quantitative**: Enter the range of the height data, **F1:F41**.
- **X/Explanatory variables: Quantitative**: Enter the range of the shoe print data, **D1:D41**.
- **Sheet**: The output will be placed in a new Excel worksheet.
- **Variable labels**: Because the top cells in the data ranges are labels and not data values, you need to select **Variable labels**.

6. Click the **Options** tab. Be sure that the selection for the confidence interval is **95%**.

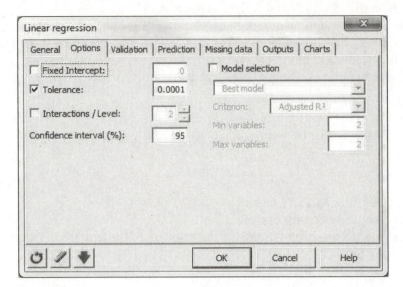

7. Click the **Prediction** tab. Under X/Explanatory variables: Quantitative, enter **G2**, the cell address of 29. Recall that you are predicting height for a shoe print length equal to 29 cm.

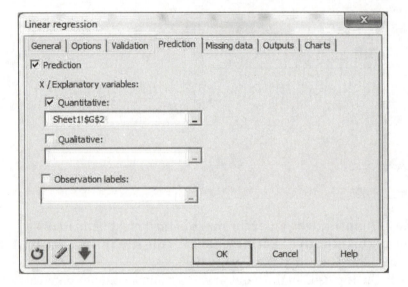

8. Click the **Outputs** tab. **Select Descriptive statistics** and **Predictions and residuals**.

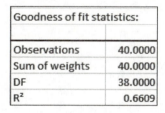

9. To reduce the amount of output, you might want to remove the check marks from any items selected on the Charts dialog box.
10. Click **OK**. Click **Continue** in the XLSTAT-Selections dialog box.

The XLSTAT output labeled *Goodness of fit statistics* is shown below. This output includes the coefficient of determination, R^2. For this analysis, the coefficient of determination is equal to 0.661.

Goodness of fit statistics:	
Observations	40.0000
Sum of weights	40.0000
DF	38.0000
R^2	0.6609

The XLSTAT output also includes the prediction equation.

Equation of the model:		
Height = 80.93041+3.21856*Shoe Print		

Further down in the output, you will find the predicted height for a shoe print length of 29 cm. The predicted height is 174.3 cm.

Predictions for the new observations:		
Observation	Pred(Height)	Std. dev. on pred. (Mean)
PredObs1	174.2687	0.9398

Scroll to the right in the XLSTAT output to see the lower and upper bounds of the 95% prediction interval. The lower bound is equal to 162.1 and the upper bound is equal to 186.5.

Lower bound 95% (Observation)	Upper bound 95% (Observation)
162.0867	186.4507

10-5 Multiple Regression

Exercise 13, page 549	**Using XLSTAT to Carry Out a Multiple Regression Analysis to Predict Amount of Nicotine in Cigarettes**

In this section, you will learn how to carry out a multiple regression analysis using XLSTAT. To illustrate multiple regression, we will use the CIGARET data file on your data disk. These data are referenced in Exercise 13 on page 549 in the Triola text. The instructions ask you to find the best regression equation for predicting the amount of nicotine in cigarettes that are 100 mm long, filtered, nonmenthol, and nonlight.

1. Begin by opening the **CIGARET** data file on your data disk. The first few rows of the data set are shown below. The tar (FLTar), nicotine (FLNic), and carbon monoxide (FLCO) variables we will be using are located in columns G, H, and I, respectively.

	A	B	C	D	E	F	G	H	I
1	KgTar	KgNic	KgCO	MnTar	MnNic	MnCO	FLTar	FLNic	FLCO
2	20	1.1	16	16	1.1	15	5	0.4	4
3	27	1.7	16	13	0.8	17	16	1	19
4	27	1.7	16	16	1	19	17	1.2	17

2. The dependent (*Y*) variable for the analysis is FLNic, and the explanatory (*X*) variables are FLTar and FLCO. The explanatory variables need to be in adjacent columns in the worksheet. I copied the three variables to sheet 2 and rearranged them so that the explanatory variables, FLTar and FLCO, were in adjacent columns as shown below.

	A	B	C
1	FLNic	FLTar	FLCO
2	0.4	5	4
3	1	16	19
4	1.2	17	17

3. At the top of the screen, select the **XLSTAT** add-in.
4. Select **Modeling data**. Select **Linear regression**.

5. Complete the dialog box as shown below. A description of the entries is given immediately after the dialog box.

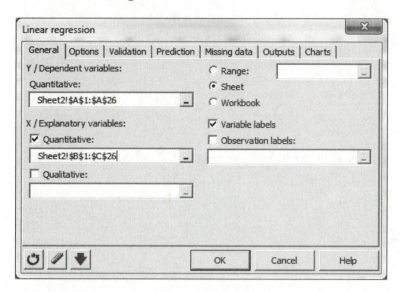

- **Y/Dependent variables**: **Quantitative**: Enter FLNic range, **A1:A26**.
- **X/Explanatory variables: Quantitative**: Enter the range of FLTar and FLCO, **B1:C26**.
- **Sheet**: The output will be placed in a new Excel worksheet.
- **Variable labels**: A check mark in the box next to **Variable labels** lets XLSTAT know that the top cell in each column is a label and not a data value.

6. Click the **Options tab** at the top of the dialog box. We will use the default value of **95%** for confidence intervals.
7. You will not be requesting any procedures related to validation, prediction, or charts. Because there are no missing data in the CIGARET data file, you do not need to select a procedure for dealing with missing data. Click the **Outputs** tab. Select the first three output options: **Descriptive statistics**, **Correlations**, and **Analysis of variance**.
8. Click **OK**. Click **Continue** at the bottom of the XLSTAT-Selections dialog box.

The XLSTAT output is shown below and on the next page. The correlation matrix displays the Pearson zero-order correlation between each pair of variables. The zero-order correlation between FLNIC and FLTar is 0.939, and the zero-order correlation between FLNic and FLCO is 0.679.

Correlation matrix:			
Variables	FLTar	FLCO	FLNic
FLTar	**1.0000**	0.8493	0.9391
FLCO	0.8493	**1.0000**	0.6785
FLNic	0.9391	0.6785	**1.0000**

R^2 and adjusted R^2 are displayed in the XLSTAT output in a section labeled *Goodness of fit statistics.* R^2 is equal to 0.933, and adjusted R^2 is equal to 0.927.

Goodness of fit statistics:	
Observations	25.0000
Sum of weights	25.0000
DF	22.0000
R^2	0.9328
Adjusted R^2	0.9267

The multiple regression equation is displayed in the XLSTAT output in a section labeled Equation of the model.

Equation of the model:		
FLNic = 0.12714+0.08780*FLTar-0.02500*FLCO		

11

Goodness-of-Fit and Contingency Tables

11-2   Goodness-of-Fit

11-3   Contingency Tables

11-2 Goodness-of-Fit

<table>
<tr><td>Example 1, page 568</td><td>**Using the CHISQ.TEST Function to Carry Out a Goodness-of-Fit Test on the Last Digits of Weights**</td></tr>
</table>

Example 1 asks you to test the claim that the sample of last digits of weights is from a population in which the last digits do *not* occur with the same frequency. You will be using the CHISQ.TEST function to carry out a goodness-of-fit test on the data. The output is the *p*-value of observed chi-square.

1. Begin by entering the observed and expected frequencies in an Excel worksheet as shown below. Click in cell **D1** where the output will be placed.

	A	B	C	D
1	Observed	Expected		
2	46	10		
3	1	10		
4	2	10		
5	3	10		
6	3	10		
7	30	10		
8	4	10		
9	0	10		
10	8	10		
11	3	10		

2. At the top of the screen, click **FORMULAS**. Select **Insert Function**.

3. Select the **Statistical** category. Select the **CHISQ.TEST** function.

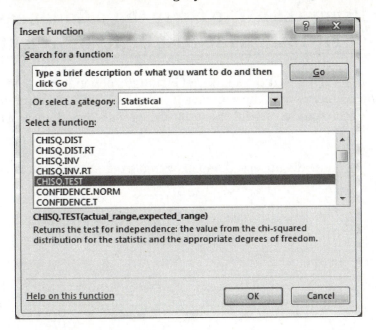

4. Complete the dialog box as shown below. A description of the entries is provided immediately after the dialog box.

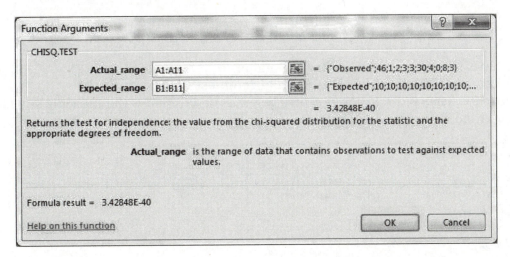

*   **Actual_range**: Enter the range of the observed frequencies, **A1:A11**.
*   **Expected_range**: Enter the range of the expected frequencies, **B1:B11**.

5. Click **OK**.

Excel's CHISQ.TEST function returns 3.42848E-40. Because the p-value is very small, it is shown in scientific notation.

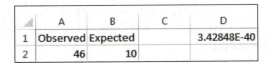

Example 1, page 568	Using XLSTAT to Carry Out a Goodness-of-Fit Test on the Last Digits of Weights

This time, you will use XLSTAT to test the claim that the sample of last digits of weights is from a population in which the last digits do *not* occur with the same frequency. The XLSTAT output for the goodness-of-fit test includes observed chi-square, critical chi-square, degrees of freedom, and the *p*-value associated with observed chi-square.

1. If you have not already done so, enter the observed and expected frequencies in an Excel worksheet as shown below.

	A	B
1	Observed	Expected
2	46	10
3	1	10
4	2	10
5	3	10
6	3	10
7	30	10
8	4	10
9	0	10
10	8	10
11	3	10

2. At the top of the screen, select the **XLSTAT** add-in.
3. Select **Parametric tests**. Select **Multinomial goodness of fit test**.
4. Complete the General dialog box as shown below. A description of the entries is provided right after the dialog box.

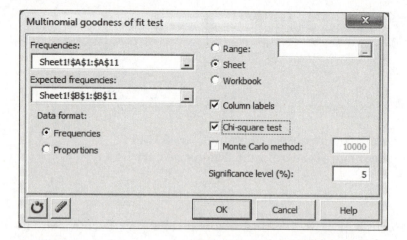

- **Frequencies**: Enter the range of the observed frequencies, **A1:A11**.
- **Expected frequencies**: Enter the range of the expected frequencies, **B1:B11**.
- **Data format**: You can enter either expected proportions or expected frequencies. Because you entered expected frequencies, select **Frequencies**.
- **Sheet**. The output will be placed in a new Excel worksheet.

- **Column labels**: The worksheet ranges you entered for the observed and expected frequencies include a label in the top cell. You need to check **Column labels** so that the XLSTAT procedure will treat the contents of these cells as labels and not as data values.
- **Chi-square test**: Select **Chi-square test** so that the output will include the results of a chi-square goodness-of-fit test.
- **Significance level (%)**: Use the default significance level of **5%**.

5. Click **OK**. Click **Continue** in the XLSTAT-Selections dialog box.

The XLSTAT output is shown below. The observed chi-square is equal to 212.80, the critical chi-square is equal to 16.92, degrees of freedom are equal to 9, the *p*-value is less than 0.0001, and the alpha (significance level) is 0.05. The test interpretation tells you that the result is statistically significant. You would conclude that last digits do not occur with the same frequency.

Chi-square test:					
Chi-square (Observed value)	212.8000				
Chi-square (Critical value)	16.9190				
DF	9				
p-value	< 0.0001				
alpha	0.05				
Test interpretation:					
H0: The distribution is not different from what is expected.					
Ha: The distribution is different from what is expected.					
As the computed p-value is lower than the significance level alpha=0.05, one should reject the null hypothesis H0, and accept the alternative hypothesis Ha.					
The risk to reject the null hypothesis H0 while it is true is lower than 0.01%.					

11-3 Contingency Tables

Example 1, page 577

Using XLSTAT to Carry Out a Chi-Square Test of Independence

To carry out a chi-square test of independence using XLSTAT, you can start with either the observed frequency table or the data. The instructions that are given here assume that you have constructed the observed frequency table. The observed frequency table, Table 11-6, is found in Example 1 of Section 11-3 on page 577 of the Triola text.

1. Begin by entering the observed frequency table in an Excel worksheet as shown at the top of the next page. I used abbreviations for the row labels.

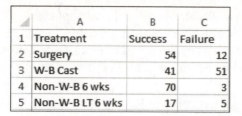

	A	B	C
1	Treatment	Success	Failure
2	Surgery	54	12
3	W-B Cast	41	51
4	Non-W-B 6 wks	70	3
5	Non-W-B LT 6 wks	17	5

2. At the top of the screen, select the **XLSTAT** add-in.
3. Select **Correlation/Association tests**. Select **Tests on contingency tables (Chi-square)**.
4. Complete the General dialog box as shown below. A description of the entries is given immediately after the dialog box.

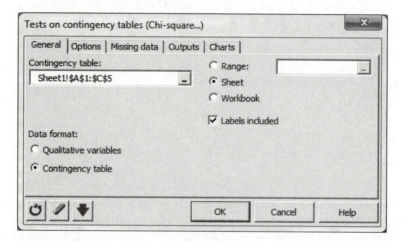

- **Contingency table**: Enter the range of the observed frequency table, **A1:C5**.
- **Data format**: Select **Contingency table**.
- **Sheet**: The output will be placed in a new Excel worksheet.
- **Labels included**: Because the row and column labels are included in the contingency table range, you need to check **Labels included**.

5. Click the **Options tab**. Select the **Chi-square test**. Set the significance level equal to **5%**.

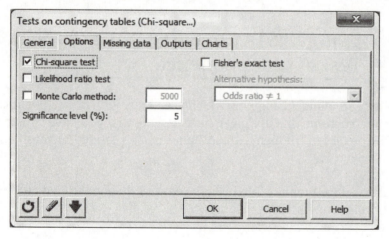

6. There are no missing data, so you don't have to enter anything in the Missing Data dialog box. You also don't have to enter anything in Charts. However, you will want to specify what you would like to be included in the output. Click the **Outputs** tab. Select **Contingency table**, and select **Association coefficients**.

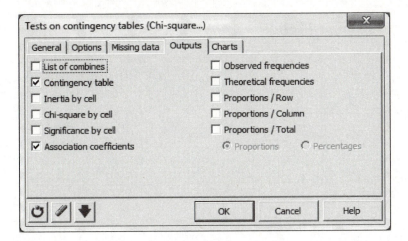

7. Click **OK**. Click **Continue** in the XLSTAT-Selections dialog box.

The XLSTAT output for the chi-square test of independence is shown below. The output includes the test results and test interpretation as well as a number of association coefficients that are not shown here.

Test of independence between the rows and the columns (Chi-square):			
Chi-square (Observed value)	58.3933		
Chi-square (Critical value)	7.8147		
DF	3		
p-value	< 0.0001		
alpha	0.05		
Test interpretation:			
H0: The rows and the columns of the table are independent.			
Ha: There is a link between the rows and the columns of the table.			
As the computed p-value is lower than the significance level alpha=0.05, one should reject the null hypothesis H0, and accept the alternative hypothesis Ha.			
The risk to reject the null hypothesis H0 while it is true is lower than 0.01%.			

12

Analysis of Variance

12-2   One-Way ANOVA

Example 1, page 602

Using Data Analysis Tools to Carry Out a One-Way ANOVA on IQ's of Three Lead Level Groups

Example 1 asks you to test the claim that the three lead level samples come from populations with means that are all equal. You are to use a significance level of $\alpha = 0.05$. The dependent variable is performance IQ and the independent variable is lead level. The low, medium, and high lead level groups are designated as 1, 2, and 3, respectively in the IQLEAD data set on your data disk. The output from Excel's Data Analysis Tools includes descriptive statistics and an analysis-of-variance table. Pairwise comparisons are not available.

1. Begin by opening the **IQLEAD** data file. Copy the LEAD and IQP variables to sheet 2 and rearrange the data so that the data for each lead level group is in a separate column. I placed Low in column A, Medium in column B, and High in column C. The first few rows are shown below.

	A	B	C
1	Low	Medium	High
2	85	78	93
3	90	97	100
4	107	107	97
5	85	80	79

2. Click **DATA** at the top of the screen and select **Data Analysis** on the far right of the data ribbon.

If Data Analysis does not appear as a choice in the data ribbon, you will need to load the Microsoft Excel ToolPak add-in. Follow the procedure on page 10.

3. Select **Anova: Single Factor**. Click **OK**.

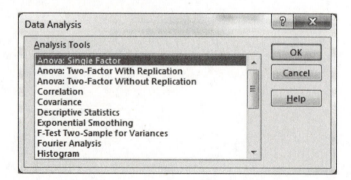

4. Complete the dialog box as shown at the top of the next page. A description of the entries is given immediately after the dialog box.

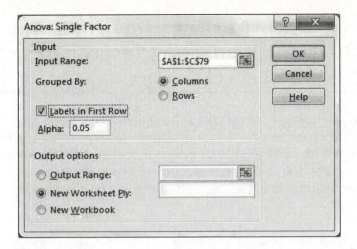

- **Input Range:** Enter the data range that includes all three lead groups, **A1:C79**.
- **Grouped By:** The data are grouped by **Columns**.
- **Labels in First Row:** Because the top cells in the range contain labels, you need to select **Labels in First Row**.
- **Alpha:** Set alpha equal to **0.05** for the test.
- **Output options:** Select **New Worksheet Ply** to place the output in a new Excel worksheet.

5. Click **OK**.

The Excel output for one-way ANOVA is shown below. The output includes descriptive statistics for the three lead level groups and an ANOVA table.

Anova: Single Factor						
SUMMARY						
Groups	Count	Sum	Average	Variance		
Low	78	8011	102.7051	281.795		
Medium	22	2071	94.13636	239.5519		
High	21	1978	94.19048	129.2619		
ANOVA						
Source of Variation	SS	df	MS	F	P-value	F crit
Between Groups	2022.73	2	1011.365	4.071122	0.01951	3.07309
Within Groups	29314.05	118	248.4241			
Total	31336.78	120				

Example 1 and Example 2, pages 602, 608

Using XLSTAT to Carry Out a One-Way ANOVA and Bonferroni Tests on IQ's of Three Lead Level Groups

You will be using XLSTAT to carry out a one-way ANOVA on the performance IQ data referenced in Example 1 on page 602 of the Triola text. You are testing the claim that the low, medium, and high lead level samples come from populations with performance IQ means that are equal. Because XLSTAT's one-way ANOVA procedure includes optional pairwise comparisons, we will be conducting the overall F-test and pairwise comparisons using the Bonferroni test described in Example 2 on page 608 of the Triola text.

1. Begin by opening the **IQLEAD** data file on your data disk. The first few rows are shown below. The independent variable for the analysis is LEAD. The low, medium, and high lead levels are designated as 1, 2, and 3, respectively. The dependent variable is IQP, performance IQ.

	A	B	C	D	E	F	G	H
1	LEAD	AGE	SEX	YEAR1	YEAR2	IQV	IQP	IQF
2	1	11	1	25	18	61	85	70
3	1	9	1	31	28	82	90	85
4	1	11	1	30	29	70	107	86
5	1	6	1	29	30	72	85	76

2. At the top of the screen, select the **XLSTAT** add-in.
3. Select **Modeling data**. Select **ANOVA**.
4. Complete the General dialog box as shown below. A description of the entries is given immediately after the dialog box.

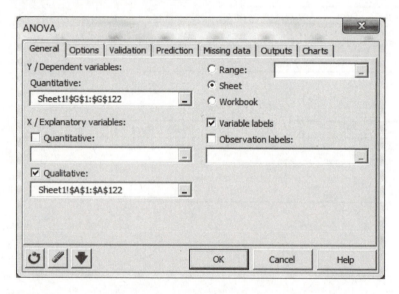

- **Y/Dependent variables: Quantitative**: Enter the range of IQP, **G1:G122**.
- **X/Explanatory variables: Qualitative**: Enter the range of LEAD, **A1:A122**.
- **Sheet**: The output will be placed in a new Excel worksheet.

- **Variable labels**: The checkmark here lets XLSTAT know that the top cells in the variable ranges are labels and not data values.

5. Click the **Options** tab. Select **95%** for the confidence interval in order to set alpha equal to 0.05 for the overall *F*-test. Nothing else needs to be selected in this dialog box.

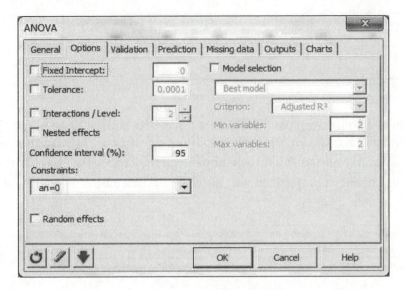

6. You will not select anything in the Validation or Prediction dialog boxes. Because there are no missing data, you can skip the Missing Data dialog box. Click the **Outputs** tab. Select **Descriptive statistics**, **Analysis of variance**, and **Pairwise comparisons**. Select the **Bonferroni** procedure.

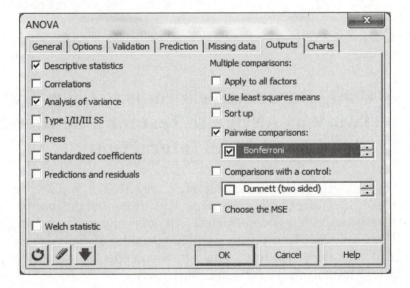

7. Click **OK**. Click **Continue** in the XLSTAT-Selections dialog box.

The XLSTAT output appears in an Excel worksheet labeled *ANOVA*. The analysis-of-variance table is shown below. The observed *F* is equal to 4.071 and the *p*-value associated with observed *F* is 0.0195. Because the *p*-value is less than alpha (0.05), the result is statistically significant.

Analysis of variance:					
Source	DF	Sum of squares	Mean squares	F	Pr > F
Model	2	2022.7299	1011.3650	4.0711	0.0195
Error	118	29314.0470	248.4241		
Corrected	120	31336.7769			
Computed against model Y=Mean(Y)					

Scroll down to find the results of the Bonferroni pairwise comparison procedure. None of the three pairwise comparisons are statistically significant. Note the modified significance level at the bottom of the table. Alpha is distributed equally across the three comparisons, $0.05/3 = 0.0167$. This tells you that the statistical significance of each comparison was determined by comparing the *p*-value of the obtained result with 0.0167.

LEAD / Bonferroni / Analysis of the differences between the categories with a confidence interval of 95%:						
Contrast	Difference	Standardized difference	Critical value	Pr > Diff	Significant	
1 vs 2	8.5688	2.2521	2.4286	0.0262	No	
1 vs 3	8.5147	2.1974	2.4286	0.0299	No	
3 vs 2	0.0541	0.0113	2.4286	0.9910	No	
Modified significance level:			0.0167			

12-3 Two-Way ANOVA

Example 1, page 618

Using Data Analysis Tools to Carry Out a Two-Way ANOVA to Test for a Gender-by-Blood-Lead-Level Interaction

To illustrate how to use Excel's Data Analysis Tools to carry out a two-way ANOVA, you will be entering the data displayed in Table 12-3 on page 615 of the Triola text. Note that the number of observations in each blood lead level group is the same. The test will not perform correctly if group sizes are unequal. Also, the Excel two-way ANOVA procedure will not produce a result if there are any missing data. Example 1 instructions ask you to test for an interaction effect, a main effect of gender, and a main effect of blood lead level using a 0.05 significance level.

1. Open a new Excel worksheet and enter the Table12-3 data as shown at the top of the next page.

	A	B	C	D
1	Gender	Low	Medium	High
2	Male	85	78	93
3		90	107	97
4		107	90	79
5		85	83	97
6		100	101	111
7	Female	64	97	100
8		111	80	71
9		76	108	99
10		136	110	85
11		99	97	93

2. Click **Data** at the top of the screen. Click **Data Analysis** on the right side of the data ribbon.

If Data Analysis does not appear as a choice in the data ribbon, you will need to load the Microsoft Excel ToolPak add-in. Follow the procedure on page 10.

3. Select **Anova: Two-Factor With Replication** and click **OK**.

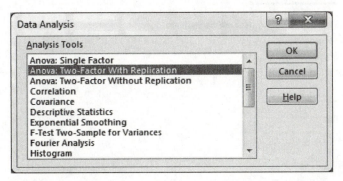

4. Complete the dialog box as shown below. A description of the entries is given immediately after the dialog box.

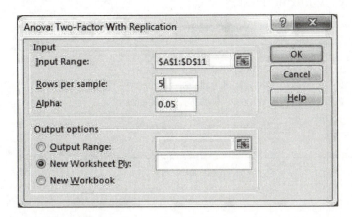

- **Input Range**: Enter the worksheet range of the data table, **A1:D11**.
- **Rows per sample**: There must be the same number of rows in each sample. For this problem, the first five rows contain the males' data and the next five rows contain the females' data.

- **Alpha**: The significance level is set equal to **0.05**.
- **Output options**: The output will be placed in a new worksheet.

5. Click **OK**.

Excel's two-way ANOVA output includes a summary section and an analysis-of-variance table. The summary section provides descriptive statistics for males, for females, and for males and females combined.

◢	A	B	C	D	E
1	Anova: Two-Factor With Replication				
2					
3	SUMMARY	Low	Medium	High	Total
4	*Male*				
5	Count	5	5	5	15
6	Sum	467	459	477	1403
7	Average	93.4	91.8	95.4	93.53333
8	Variance	95.3	146.7	130.8	108.8381
9					
10	*Female*				
11	Count	5	5	5	15
12	Sum	486	492	448	1426
13	Average	97.2	98.4	89.6	95.06667
14	Variance	812.7	142.3	143.8	330.2095
15					
16	*Total*				
17	Count	10	10	10	
18	Sum	953	951	925	
19	Average	95.3	95.1	92.5	
20	Variance	407.5667	140.5444	131.3889	

In the ANOVA table, Sample is the row factor (gender). The column factor is blood lead level. The interaction is the gender × blood lead level interaction. None of the three *p*-values are less than 0.05. Therefore, there are no statistically significant effects.

23	ANOVA						
24	*Source of Variation*	*SS*	*df*	*MS*	*F*	*P-value*	*F crit*
25	Sample	17.63333	1	17.63333	0.071895	0.790889	4.259677
26	Columns	48.8	2	24.4	0.099484	0.905676	3.402826
27	Interaction	211.4667	2	105.7333	0.431095	0.65473	3.402826
28	Within	5886.4	24	245.2667			
29							
30	Total	6164.3	29				

13

Nonparametric Statistics

13-2 Sign Test

<table>
<tr><td>**Example 2, page 641**</td><td>**Using XLSTAT to Carry Out a Sign Test on Taxi-Out and Taxi-In Times**</td></tr>
</table>

You will be using XLSTAT to carry out a sign test on the flight data displayed in Table 13-3 on page 641 of the Triola text. The instructions in Example 2 ask you to use the sign test to test the claim that there is no difference between taxi-out times and taxi-in times using a significance level equal to 0.05.

1. Enter the Table 13-3 data in an Excel worksheet as shown below.

	A	B
1	Taxi Out	Taxi In
2	13	13
3	20	4
4	12	6
5	17	21
6	35	29
7	19	5
8	22	27
9	43	9
10	49	12
11	45	7
12	13	36
13	23	12

2. At the top of the screen, select the **XLSTAT** add-in.
3. Select **Nonparametric tests**. Select **Comparison of two samples (Wilcoxon, Mann-Whitney, ...)**.
4. Complete the General dialog box as shown below. A description of the entries is given immediately after the dialog box.

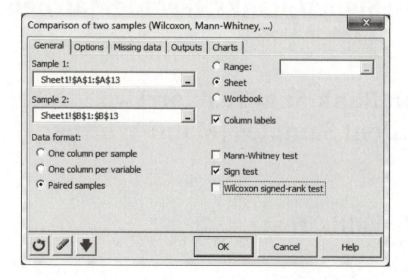

- **Sample 1:** Enter the range of the taxi-out data, **A1:A13**.
- **Sample 2:** Enter the range of the taxi-in data, **B1:B13**.
- **Data format:** Select **Paired Samples.**
- **Sheet:** The output will be placed in a new Excel worksheet.
- **Column labels:** Put a check mark here so that XLSTAT will treat the top cells in the sample ranges as column labels and not data values.
- **Sign test:** XLSTAT provides three analysis procedures. Select **Sign test**.

5. Click the **Options** tab. Complete the Options dialog box as shown below. Information about the entries is given immediately after the dialog box.

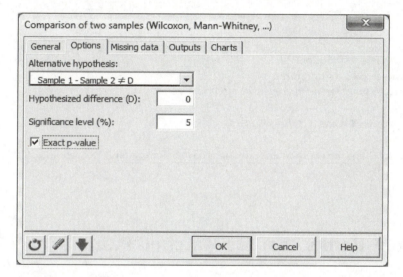

- **Alternative hypothesis:** Select the two-tailed alternative hypothesis, **Sample 1 – Sample 2 ≠ D**.
- **Hypothesized difference (D):** Select **0**, the default value.
- **Significance level (%):** Select **5%**, the default value.
- **Exact p-value:** You are given a choice between a continuity correction or the exact *p*-value. Select **Exact p-value**.

6. There are no missing data so you can skip the Missing Data tab. You will not select anything in the Charts dialog box. Click the **Outputs** tab. Select **Descriptive statistics**.
7. Click **OK**. Click **Continue** in the XLSTAT-Selections dialog box.

The first part of XLSTAT's sign test output presents summary statistics. These statistics include the number of observations, observations with missing data, observations without missing data, minimum, maximum, mean, and standard deviation.

Summary statistics:							
Variable	Observations	Obs. with missing data	Obs. without missing data	Minimum	Maximum	Mean	Std. deviation
Taxi Out	12	0	12	12.0000	49.0000	25.9167	13.4331
Taxi In	12	0	12	4.0000	36.0000	15.0833	10.6212

The next part of XLSTAT's sign test output, shown below, presents the results of the sign test and the test interpretation. Based on the result, you would not reject the null hypothesis. You would conclude that the two samples follow the same distribution.

Sign test / Two-tailed test:					
N+	8				
Expected value	5.5000				
Variance (N+)	2.7500				
p-value (Two-tailed)	0.2266				
alpha	0.05				
The p-value is computed using an exact method.					
Test interpretation:					
H0: The two samples follow the same distribution.					
Ha: The distributions of the two samples are different.					
As the computed p-value is greater than the significance level alpha=0.05, one cannot reject the null hypothesis H0.					
The risk to reject the null hypothesis H0 while it is true is 22.66%.					
Ties have been detected in the data and the appropriate corrections have been applied.					

13-3   Wilcoxon Signed-Ranks Test for Matched Pairs

Example 1, page 651	Using XLSTAT to Carry Out a Wilcoxon Signed-Ranks Test for Matched Pairs

Example 1 asks you to test the claim that there is no difference between taxi-out times and taxi-in times of American Airlines Flight 21. You will use XLSTAT to carry out a Wilcoxon signed-ranks test with a 0.05 significance level.

1. Enter the data in an Excel worksheet as shown at the top of the next page. You may have already entered these data if you completed Example 2 on page 641 of the Triola text.

	A	B
1	Taxi Out	Taxi In
2	13	13
3	20	4
4	12	6
5	17	21
6	35	29
7	19	5
8	22	27
9	43	9
10	49	12
11	45	7
12	13	36
13	23	12

2. At the top of the screen, select the **XLSTAT** add-in.
3. Select **Nonparametric tests**. Select **Comparison of two samples (Wilcoxon, Mann-Whitney, ...)**.
4. Complete the General dialog box as shown below. A description of the entries is given immediately after the dialog box.

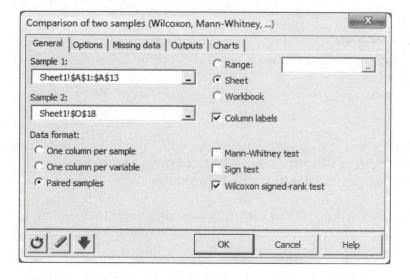

- **Sample 1:** Enter the range of the taxi-out data, **A1:A13**.
- **Sample 2:** Enter the range of the taxi-in data, **B1:B13**.
- **Data format:** Select **Paired Samples**.
- **Sheet:** The output will be placed in a new Excel worksheet.
- **Column labels:** Put a check mark here so that XLSTAT will treat the top cells in the sample ranges as column labels and not as data values.
- **Wilcoxon signed-rank test:** XLSTAT provides three options for an analysis procedure. Select **Wilcoxon signed-rank test**.

3. Next, click the **Options** tab. Complete the Options dialog box as shown below. A description of the entries is given immediately after the dialog box.

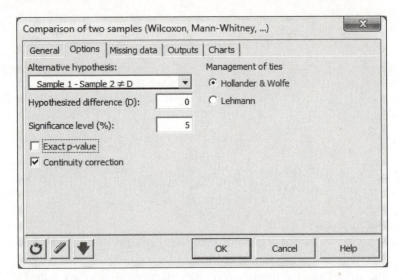

*   **Alternative hypothesis**: Select the two-tailed alternative hypothesis, **Sample 1 – Sample 2 ≠ D**.
*   **Hypothesized difference (D)**: Select **0**, the default value.
*   **Significance level (%)**: Select **5%**, the default value.
*   **Exact p-value**: You are given a choice between a continuity correction or the exact *p*-value. Select **Continuity correction**.
*   **Management of ties**: You are given two options for the management of ties. Select **Hollander & Wolfe**.

4. There are no missing data so you can skip the Missing Data tab. You will not select anything in the Charts dialog box. Click the **Outputs** tab. Select **Descriptive statistics**.
5. Click **OK**. Click **Continue** in the XLSTAT-Selections dialog box.

The first part of XLSTAT's output for the Wilcoxon signed-ranks test presents summary statistics. These statistics include the number of observations, observations with missing data, observations without missing data, minimum, maximum, mean, and standard deviation.

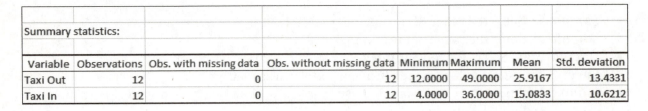

Summary statistics:							
Variable	Observations	Obs. with missing data	Obs. without missing data	Minimum	Maximum	Mean	Std. deviation
Taxi Out	12	0	12	12.0000	49.0000	25.9167	13.4331
Taxi In	12	0	12	4.0000	36.0000	15.0833	10.6212

The next part of XLSTAT's output presents the results of the Wilcoxon signed-ranks test and the test interpretation. The V in the output is the same as T in your textbook (see Step 4 on page 650 of the Triola text). V is the sum of the positive ranks and it is equal to 55. If you had entered the taxi-in data as sample 1 instead of the taxi-out data, V would have been equal to 11, the absolute sum of the negative ranks. The remaining output would be exactly the same. So it doesn't matter which sample is entered as sample 1 and which is entered as sample 2. Based on the result, you would not reject the null hypothesis. You would conclude that the two samples follow the same distribution.

Wilcoxon signed-rank test / Two-tailed test:			
V	55.0000		
Expected value	33.0000		
Variance (V)	126.3750		
p-value (Two-tailed)	0.0558		
alpha	0.05		
An approximation has been used to compute the p-value.			
Test interpretation:			
H0: The two samples follow the same distribution.			
Ha: The distributions of the two samples are different.			
As the computed p-value is greater than the significance level			
alpha=0.05, one cannot reject the null hypothesis H0.			
The risk to reject the null hypothesis H0 while it is true is 5.58%.			
Continuity correction has been applied.			
Ties have been detected in the data and the appropriate corrections have been applied.			

13-4 Wilcoxon Rank-Sum Test for Two Independent Samples (Mann-Whitney U Test)

Exercise 13, page 660 | **Using XLSTAT to Carry Out a Mann-Whitney U Test on Pulse Rates of Males and Females**

The Wilcoxon rank-sum test for two independent samples is not available in either XLSTAT or in Excel's Data Analysis Tools. However, the Mann-Whitney U test is available in XLSTAT, and it is equivalent to the Wilcoxon rank-sum test. Exercise 13 asks you to use the data in Table 13-5 on page 656 of the Triola text to find the z test statistic for the Mann-Whitney U test and to compare this value to the z test statistic found using the Wilcoxon rank-sum test ($z = -1.26$). The XLSTAT output does not provide the z statistic, but it does provide the observed U statistic and the p-value.

1. Enter the pulse rates for males and females in an Excel worksheet as shown at the top of the next page.

	A	B
1	Males	Females
2	60	78
3	74	80
4	86	68
5	54	56
6	90	76
7	80	78
8	66	78
9	68	90
10	68	96
11	56	60
12	80	98
13	62	

2. At the top of the screen, select the **XLSTAT** add-in.
3. Select **Nonparametric tests**. Select **Comparison of two samples (Wilcoxon, Mann-Whitney, ...)**.
4. Complete the General dialog box as shown below. A description of the entries is given immediately after the dialog box.

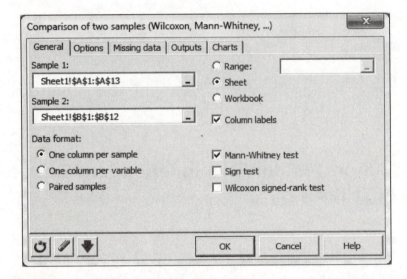

- **Sample 1**: Enter the range of the males' pulse rates, **A1:A13**.
- **Sample 2**: Enter the range of the females' pulse rates, **B1:B12**.
- **Data format**: Select **One column per sample**.
- **Sheet**: The output will be placed in a new Excel worksheet.
- **Column labels**: The check mark here lets XLSTAT know that the top cell in each range is a label and not a data value.
- **Mann-Whitney test**: Select the **Mann-Whitney test** to compare the two independent samples.

5. Click the **Options** tab. Complete the Options dialog box as shown at the top of the next page. A description of the entries immediately follows the dialog box.

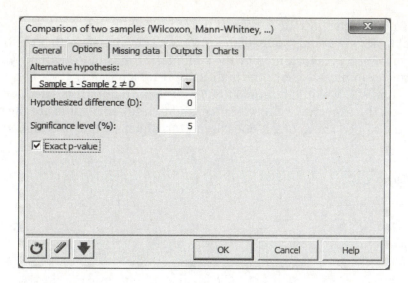

- **Alternative hypothesis**: Select the two-tailed alternative hypothesis, **Sample 1 – Sample 2 ≠ D**.
- **Hypothesized difference (D)**: Select the default value of **0**.
- **Significance level (%)**. Select the default value of **5%**.
- **Exact p-value**. Select **Exact p-value** rather than the continuity correction.

6. There are no missing data so you can skip the Missing Data tab. You can also skip the Charts tab. In the Outputs dialog box, select **Descriptive statistics**.
7. Click **OK**. Click **Continue** in the XLSTAT-Selections dialog box.

The first part of the XLSTAT output for the Mann-Whitney test includes summary statistics for the two samples.

Summary statistics:							
Variable	Observations	Obs. with missing data	Obs. without missing data	Minimum	Maximum	Mean	Std. deviation
Females	11	0	11	56.0000	98.0000	78.0000	13.3566
Males	12	0	12	54.0000	90.0000	70.3333	11.7189

The second part of the XLSTAT output displays the results of the Mann-Whitney test. For this analysis, the U statistic is equal to 45.5, whereas your textbook answer key indicates that the U statistic is equal to 86.5. The value of U is determined by the sum of the ranks in the sample you selected as sample 1, and for this example, we chose the male sample for sample 1. If we had chosen the female sample for sample 1, U would be equal to 86.5. The output for this selection is also shown below. As you can see, the expected value, variance, and two-tailed p-value are the same. The test interpretation is also the same. Therefore, it doesn't matter which sample is designated as sample 1 for the analysis.

Mann-Whitney test / Two-tailed test:		
U	45.5000	
Expected value	66.0000	
Variance (U)	262.0435	
p-value (Two-tailed)	0.2278	
alpha	0.05	
The p-value is computed using an exact method.		

Mann-Whitney test / Two-tailed test:		
U	86.5000	
Expected value	66.0000	
Variance (U)	262.0435	
p-value (Two-tailed)	0.2278	
alpha	0.05	
The p-value is computed using an exact method.		

XLSTAT's test interpretation indicates that the null hypothesis should not be rejected. The result is not statistically significant. You would conclude that the populations have the same median.

Test interpretation:						
H0: The difference of location between the samples is equal to 0.						
Ha: The difference of location between the samples is different from 0.						
As the computed p-value is greater than the significance level alpha=0.05, one cannot reject the null hypothesis H0.						
The risk to reject the null hypothesis H0 while it is true is 22.78%.						
Ties have been detected in the data and the appropriate corrections have been applied.						

13-5   Kruskal-Wallis Test

Example 1, page 663	Using XLSTAT to Carry Out a Kruskal-Wallis Test on Pulse Rates of Males and Females

The Kruskal-Wallis test is a nonparametric test available in XLSTAT. In the XLSTAT output, the test statistic is designated as K, whereas in your textbook the test statistic is designated as H. When there are ties, the Kruskal-Wallis test available in XLSTAT adjusts the result for the number of ties. Therefore, when there are ties in the data, the result you obtain using the computing formula on page 662 of the Triola text will not be exactly the same as the result you obtain when you use XLSTAT. However, the results will be very close. To illustrate how to carry out a Kruskal-Wallis test using XLSTAT, we will do the problem presented in Example 1 on page 663 of the Triola text. Example 1 asks you to use a 0.05 significance level to test the claim that the three lead level samples come from populations with medians that are all equal.

1. Enter the performance IQ scores for the low, medium, and high lead level groups in an Excel worksheet as shown below.

	A	B	C
1	Low	Medium	High
2	85	78	93
3	90	97	100
4	107	107	97
5	85	80	79
6	100	90	97
7	97	83	
8	101		
9	64		

2. At the top of the screen, select the **XLSTAT** add-in.
3. Select **Nonparametric tests**. Select **Comparison of k samples (Kruskal-Wallis, Friedman, ...)**.
4. Complete the General dialog box as shown at the top of the next page. A description of the entries immediately follows the dialog box.

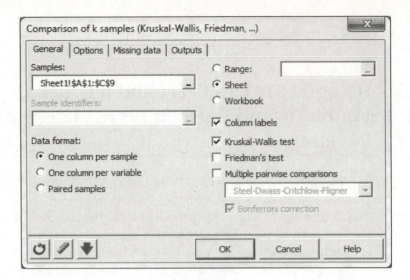

- **Samples**: Enter the range of the three samples, **A1:C9**.
- **Data format**: Select **One column per sample**.
- **Sheet**: The output will be placed in a new Excel worksheet.
- **Column labels**: Select **Column labels** to let XLSTAT know that the top cells in the columns contain labels, not data values.
- **Kruskal-Wallis test**: Select the **Kruskal-Wallis** test for an analysis of independent samples.

5. Click the **Options** tab. Select the default significance level of **5%**. Select the **Asymptotic p-value**.

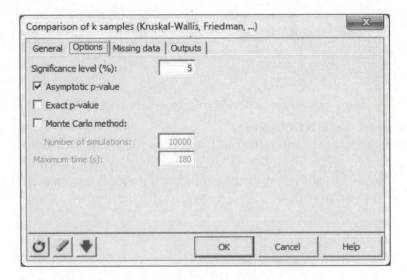

6. Click **Missing data** and select **Ignore missing data**. Click **Outputs** and select **Descriptive Statistics**.
7. Click **OK**. Click **Continue** in the XLSTAT-Selections dialog box.

The XLSTAT output contains summary statistics for each lead level group. The statistics include observations, observations with missing data, observations without missing data, minimum, maximum, mean, and standard deviation.

Summary statistics:							
Variable	Observations	Obs. with missing data	Obs. without missing data	Minimum	Maximum	Mean	Std. deviation
Low	8	0	8	64.0000	107.0000	91.1250	13.4954
Medium	8	2	6	78.0000	107.0000	89.1667	11.1967
High	8	3	5	79.0000	100.0000	93.2000	8.3187

The XLSTAT output also contains test results and a test interpretation. The observed K is equal to 0.7031, whereas in your text, observed H is equal to 0.694. The difference in the observed values is due to the adjustment that the XLSTAT procedure makes to the Kruskal-Wallis test when there are ties. The last line of the output lets you know that this adjustment has been made.

Kruskal-Wallis test:			
K (Observed value)	0.7031		
K (Critical value)	5.9915		
DF	2		
p-value (Two-tailed)	0.7036		
alpha	0.05		
An approximation has been used to compute the p-value.			
Test interpretation:			
H0: The samples come from the same population.			
Ha: The samples do not come from the same population.			
As the computed p-value is greater than the significance level alpha=0.05, one cannot reject the null hypothesis H0.			
The risk to reject the null hypothesis H0 while it is true is 70.36%.			
Ties have been detected in the data and the appropriate corrections have been applied.			

14

Statistical Process Control

14-2   Control Charts for Variation and Mean

Exercise 5, page 701	Using XLSTAT to Find the Values of LCL and UCL for an R Chart and an X-Bar Chart

You will be using XLSTAT to produce control charts for variation and mean. Exercise 5 asks you to find the sample mean and range ($\bar{x}$ and $\bar{R}$) for each of the 25 days. You are also to find values of LCL and UCL for an R chart, and then the values of LCL and UCL for an $\bar{x}$ chart. You will be using Data Set 22 in Appendix B. These data are also available in the CANS file on your data disk. The data are axial loads of aluminum cans that are 0.0109 in. thick. There are seven measurements from each of 25 days of production.

1. Open the **CANS** data file on your data disk and rearrange the data so that it appears in the same way as Data Set 22 on page 763 of the Triola text. The first few lines are shown below.

	A	B	C	D	E	F	G	H
1	1	270	273	258	204	254	228	282
2	2	278	201	264	265	223	274	230
3	3	250	275	281	271	263	277	275
4	4	278	260	262	273	274	286	236
5	5	290	286	278	283	262	277	295

2. At the top of the screen, select the **XLSTAT** add-in.
3. Select **SPC**. Select **Subgroup charts**.
4. Click the **Mode** tab at the top. For chart family, select **Subgroup charts**. For chart type, select **R chart**.

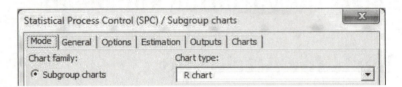

5. Click the **General** tab. Complete the dialog box as shown at the top of the next page. Note that the data range, **B1:H25**, does *not* include the sample numbers in column A. Also be sure there is no check mark in the box next to **Column labels**.

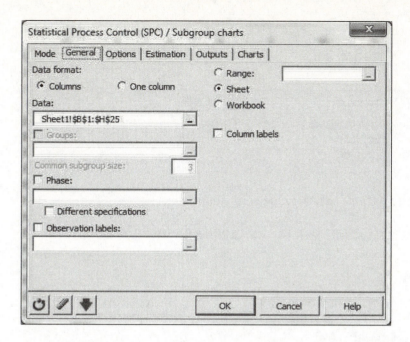

6. You will not be selecting anything in Options. Click the **Estimation** tab. Select **R bar**.

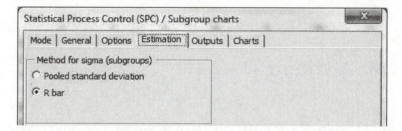

7. In the **Outputs** and **Charts** dialog boxes, you can use the default selections.
8. Click **OK**. Click **Continue** in the XLSTAT-Selections dialog box.

The XLSTAT output for the R chart is shown below. The mean is equal to 267.11, and the standard deviation is equal to 20.32. For the R chart, UCL is 105.75 and LCL is 4.17.

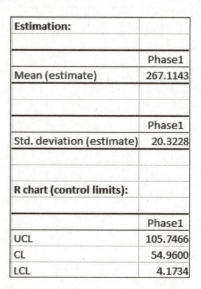

Estimation:	
	Phase1
Mean (estimate)	267.1143
	Phase1
Std. deviation (estimate)	20.3228
R chart (control limits):	
	Phase1
UCL	105.7466
CL	54.9600
LCL	4.1734

9. Next, you will find the values for the $\bar{x}$ chart. Click **SPC** at the top of the screen and select **Subgroup charts**.

10. Click the **Mode** tab, and select **X-bar chart**.

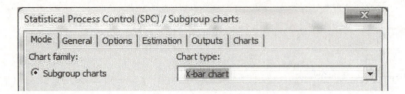

11. Click **OK**. Click **Continue** in the XLSTAT-Selections dialog box.

The XLSTAT output for the $\bar{x}$ chart is shown below. UCL is 290.16 and LCL is 244.07.

X-bar chart (control limits):		
	Phase1	
UCL	290.1581	
CL	267.1143	
LCL	244.0704	

14-3   Control Charts for Attributes

Example 1, page 703	**Using XLSTAT to Construct a Control Chart for the Proportion of Defective Quarters**

Example 1 asks you to construct a control chart for the proportion p of defective quarters and to determine whether the process is within statistical control. The data represent the numbers of defects in batches of 10,000 quarters randomly selected on each of 20 consecutive days from a new manufacturing process that is being tested.

1. Enter the data in an Excel worksheet as shown at the top of the next page.

◢	A
1	Defects
2	8
3	7
4	12
5	9
6	6
7	10
8	10
9	5
10	15
11	14
12	12
13	14
14	9
15	6
16	16
17	18
18	20
19	19
20	18
21	24

2. At the top of the screen, select the **XLSTAT** add-in.
3. Select **SPC**. Select **Attribute charts**.
4. Complete the **Mode** dialog box as shown below. For chart family, select **Attribute charts**. For chart type, select **P chart**.

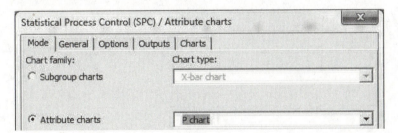

5. Click the **General** tab. Complete the dialog box as shown at the top of the next page. A description of the entries immediately follows the dialog box.

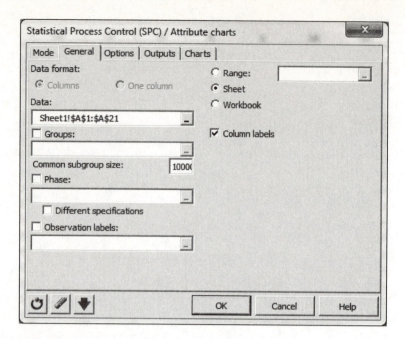

- **Data:** Enter the data range, **A1:A21**.
- **Common subgroup size:** The common subgroup size is **10000**.
- **Sheet:** The output will be placed in a new Excel worksheet.
- **Column labels:** Because the top cell in the data range is a label and not a data value, you need to select **Column labels**.

6. Click the **Charts** tab. Use the default selections as shown below.

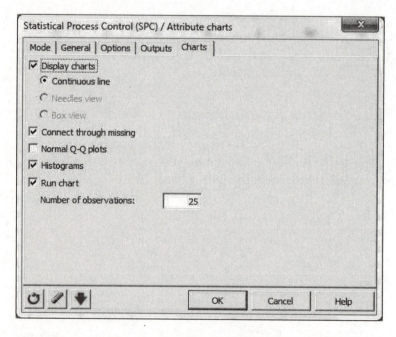

7. Click **OK**. Click **Continue** in the XLSTAT-Selections dialog box.

The XLSTAT output is very extensive and only some of the output is shown here. The Estimation section of the output displays the mean (estimate) (0.0013). The P chart (control limits) section displays the values of UCL (0.0023), and LCL (0.0002).

Estimation:	
	Phase1
Mean (estimate)	0.0013
P chart (control limits):	
	Phase1
UCL	0.0023
CL	0.0013
LCL	0.0002

A P chart is also included in the XLSTAT output.

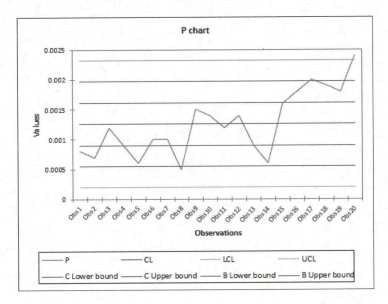